Michigan APPLES

Michigan APPLES
HISTORY & TRADITION

SHARON KEGERREIS
Foreword by Sharon Steffens

AMERICAN PALATE

Published by American Palate
A Division of The History Press
Charleston, SC 29403
www.historypress.net

Copyright © 2015 by Sharon Kegerreis
All rights reserved

Front cover, top: Janet Kline, Gladys Davenport and Alma Gillett load up Apple Queen Carol Fahling's serving box at the 1953 Apple Smorgasbord. *Courtesy of Alpine Township Historical Commission.*

Front cover, center: Ginger Gold apple. *Photo by Lorri Hathaway.*

All photographs are from the author's collection unless otherwise noted.

First published 2015

ISBN 978-1-5402-1020-3

Library of Congress Control Number: 2014954497

Notice: The information in this book is true and complete to the best of our knowledge. It is offered without guarantee on the part of the author or The History Press. The author and The History Press disclaim all liability in connection with the use of this book.

All rights reserved. No part of this book may be reproduced or transmitted in any form whatsoever without prior written permission from the publisher except in the case of brief quotations embodied in critical articles and reviews.

*To Kris, who planted apple trees and encouraged me to write about apples.
To Julia, who was the reason for my first cider mill adventure fifteen years ago
and who checked out massive research tomes for me at her new college digs.
To Makayla, who joined me on many apple outings and was an excellent taste tester.
I love you all.*

Contents

Foreword, by Sharon Steffens	9
Introduction	11
1. Veritable Paradise of Fruits	15
2. Markets Bloom	37
3. Apples to Apples	55
4. Four Seasons	63
5. Into the Orchards	77
6. Tastes of History	89
7. The Hunt for Sweet October	101
8. The Intoxicating Apple	113
Epilogue. Plant It Forward	127
Acknowledgements	129
Appendix. Glossary and Resources	133
Works Cited	137
Index	149
About the Author	157

FOREWORD

When author Sharon Kegerreis requested an interview and later wondered if I would consider writing the foreword to this book, I was intrigued. She had been researching the history of apples for some time, and because I like apple everything, I was hooked.

Teaching children about the life cycle of apples and pumpkins at our farm is a joy to me, but I hadn't really thought about how the apple tree was first introduced to Michigan. I assumed that Johnny Appleseed had something to do with it—not so. You will be surprised by the rich agricultural history of Detroit and southeast Michigan. It is amazing how the trail of apple seeds across our state and our nation has impacted the quality of everyday life and the economy of Michigan.

This book reflects hours and months of research and presents the story of apples, growers and this constantly changing industry in many different ways. Milestones of historical events help you place the context of the changes as the fledgling apple industry spread, grew and matured.

The chapter "Four Seasons" really spoke to me. When you are married to an apple grower, your life and your family's life are greatly impacted by the cycle of growth of the apple tree. You often plan your weddings, vacations and family outings around the season and the rhythm of Mother Nature. When my husband, John, and I planned our wedding almost fifty-five years ago, I had to explain to my school superintendent why I needed a week's vacation in February to get married. I was a city gal marrying an apple farmer, and any other month would not work. The farm and the apple tree

Foreword

have a significant priority in your life because that is your vocation, which supports your income and, therefore, your lifestyle.

Raising children on a farm provides great opportunities for them to learn and develop a good work ethic that will help them throughout life. Picking up drops, 4-H projects and learning to drive a tractor are great skill builders. However, there also needs to be time to play and climb an apple tree. We still have two twenty-five-foot apple trees that John's dad planted in about 1956 in our yard. These were our children's favorite trees to climb and now our grandchildren's.

Apples are so versatile. You can use them in almost everything you cook, bake or enjoy fresh. I have a friend who can't wait for the Honeycrisp to be ready each autumn so she can enjoy them with a large pail of peanut butter. Honeycrisp is the number-one apple that many look for in a farm market, but there are many other varieties that are also excellent. Customers often ask what is good for pie or sauce, and I like to encourage them to use more than one variety of apple because the blend makes it even better. It's fun to explore the taste of apples. They can be quite different in taste, smell and texture.

New varieties seem to appear on the horizon quite frequently, and as growers, we always keep an eye out for that oddity that appears in the orchard. Could it be a new, interesting, tasty and marketable apple? More than one west Michigan grower has discovered a new apple. Perhaps most notable is Paula Red, which was discovered by Luke Arends in his Sparta orchard in 1960.

The chapter "Apples to Apples" tells you the story of how apple varieties are developed and the intense scientific support system that exists to help growers and processors provide customers with the best quality product possible.

Many innovative growers have developed new markets in hard cider and apple wine. "The Intoxicating Apple" chapter tells a story of a rebirth of this aspect of the apple that is fascinating and exploding at this time.

This book is your opportunity to explore the awesomeness of the apple, the industry and the people who grow it and make it available for you to eat and enjoy.

Sharon Steffens
Steffens Orchards of Alpine and Sparta Townships (1862)

Introduction

In the backyard of my childhood home is a massive apple tree. When my parents built their home in northern Michigan in 1969, the tree's roots were deep, and its bearing was impressive. Over the years, it provided bushels of apples for pies and snacks and was the cornerstone of our homestead and harbinger of autumn. Our lips puckered at every first bite, and Mom worked hard to find the perfect balance of tart and sweet in her pies. The tree is well over one hundred years old now. Until recently, I never thought to wonder how the tree originated. Perhaps a Native American planted a seed after interacting with a Jesuit missionary? (Well, it's not *that* old.) Was a seed intentionally planted there with the hope that it would grow into a tree bearing palatable apples? Did an early pioneer purposefully plant a grafted tree to bear favorite apples like those grown on an old homestead for hard cider, apple butter or vinegar? Or did a bird drifting by drop a seed that was later trampled into the dirt by a fleeting deer? *Malus domestica*, the sweet apple, is not native to the New World. So how did the apple tree get there?

The yearlong journey to tell the Michigan apple story uncovered pioneer adventures and perseverance to grow food for families and, eventually, the commercial market. The discoveries highlight Detroit's beginnings

Introduction

and fruit farmers of the mid-1800s to early 1900s found in old Michigan State Pomological Society and Michigan State Horticultural Society journals. These journals and many more historic documents and resources revealed the apple history of Michigan. Conversations with apple farmers, industry experts and historians led to another apple farmer or historic milestone. Quickly, I became enamored of Michigan's apple story—and how quickly the year passed. The greatest discovery was the impressive depth of Michigan's apple heritage, deeply rooted long before John Chapman, aka Johnny Appleseed, traveled the Ohio Valley. Also surprising was the discovery of the varied tastes of apples and the unrelenting challenges farmers overcome to grow delicious, unblemished apples for us.

The sweet apple has been cultivated here since the 1700s, and its commercial viability has played a role in Michigan's agricultural history since the nineteenth century. The "Veritable Paradise of Fruits" and "Markets Bloom" chapters highlight this history and many fascinating notations, including that cider was the beverage of the pioneers and that President Thomas Jefferson obtained scions (thin branches) of Detroit's apple trees to graft on rootstock at his Monticello estate. Most surprisingly today, 65 to 70 percent of all Michigan apples are grown on a 158-square-mile ridge 20 miles northwest of downtown Grand Rapids. For twenty years, this region hosted an Apple Smorgasbord, an outdoor farm-hosted event featuring two hundred apple dishes, many of German and Swedish influence.

Overall, Michigan is home to 850 apple farmers and more than nine million apple trees. The state is ranked third in the United States for apple production, with an annual average crop load of twenty million bushels. Approximately 60 percent of Michigan's apples are processed into other products. *Michigan Apples* highlights this aspect of the industry in "Four Seasons," though it largely focuses on the fresh tastes of the apple.

Unequivocally, Honeycrisp has made its mark in modern times. The apple's fascinating origins are highlighted in "Apples to Apples." Beyond newer cultivated varieties are apples that connect strongly to Michigan's early apple heritage. These include Canadian and French apples called Fameuse (Snow) and Calville Blanc d'Hiver, which dominated the Detroit ribbon farms of the 1700s and early 1800s, as well as Detroit Red, which was growing prolifically along the Detroit River by the early nineteenth century. As Michigan bloomed with the onset of settlers, widely planted apple trees included Baldwin, McIntosh, Northern Spy, Golden Russet and Duchess of Oldenburg.

Several of Michigan's heritage varieties, plus many more newer apples, are grown in today's orchards, including tart Lodi, spicy sweet-tart Ginger Gold

INTRODUCTION

and flavorful and crunchy Fuji, Gala, Jonagold and Idared. The breadth of varieties is well beyond the few apples at the grocery store. Apple profiles in the "Into the Orchards" and "Tastes of History" chapters will tempt you to seek a range of tastes and try recipes tucked at the end of the chapters. Beverage artisans shared their passion for sweet cider in "The Hunt for Sweet October" and for apple libations in "The Intoxicating Apple" chapters.

Michigan Apples weaves together knowledge and experiences shared by apple growers, industry experts, sweet cider makers and libation maestros. Apple farmers growing 2,000 to 800,000 annual bushels with farming roots as deep as six generations provided insight into the dynamic industry.

Along the journey, I met intriguing Michiganders like Sharon Steffens and JoAnn Thome, two of the Fearsome Foursome on the Fruit Ridge, who picketed apple processors in the 1970s and fought for legislation to enable growers to negotiate pricing. History whiz Pat Cederholm of the Alpine Township Historical Commission revealed the vibrant history of Michigan's largest apple-growing region. Master cider maker Jim Hill of Hill Bros. Orchards in Grand Rapids showed off his squeaky-clean cider mill and delicious award-winning sweet cider. Expert grower Phil Schwallier shared his expertise on the four seasons of apple growing at his farmhouse by Schwallier's Country Basket in Sparta. Organic grower Dennis Mackey of Northern Natural Organics in Kaleva offered his perspective on growing apples organically and prompted me to dig into the legacy of Sam Cohodas, a fruit businessman and philanthropist of the mid-1900s. The Ward family of Eastman's Antique Apples in Wheeler shared background details of an array of colorful historic apples. Mike Beck of Uncle John's Cider Mill and Fruit House Winery offered insight on the collaboration of craft spirit and hard cider makers.

The Michigan apple story unfolds through the centuries and shines a spotlight on how families adapted to challenges ranging from volatile weather to global competition to save the family farm. My hope is that this book inspires you to visit the orchards from August through October and seek Michigan apples beyond the harvest season. Taste new varieties and those with intriguing histories, including Michigan natives Paula Red and Shiawassee Beauty. Try an apple that is bumpy or bruised. (You can always cook it!) Whip up apple butter like the pioneers. Enjoy sweet cider or a mug of hard cider. Make the apple the star of a savory dish or quick batch of applesauce.

Tackling this intriguing adventure to discover "all things apple" in Michigan, I took Mark Twain's advice:

Why not go out on a limb? That's where the fruit is.

Chapter 1
VERITABLE PARADISE OF FRUITS

*Honor waits, o'er all the earth,
Through endless generations,
the art that calls her harvests forth,
and feeds the expectant nations.*
—William Cullen Bryant

In the mid-1600s to 1700s, French Jesuit missionaries arrived in the Great Lakes region. These intrepid men traveled by birchbark canoe through waterways teeming with fish and traversed by foot the exotic shorelines lush with forest, vegetation and wildlife. Throughout their travels, they documented observations of the flora and fauna and their interactions with indigenous people. The missionaries were sent to this wilderness to minister to Native Americans and to cultivate allegiances between France and the native people. Along the way, they planted seeds with the hope of fruitful harvests.

Apple seedlings readily adapted to the northern clime and eventually grew to impressive height. Many of these trees were admired centuries later. And though the fruits of the trees were not of cultivated varieties, through the years, they provided food to the explorers and settlers who followed, in the form of apple butter, dried apples and sauce. The tangy apples supplemented the tastes of native, palatable chestnuts, cranberries and grapes and bitter crab apples.

Just imagine: by the end of the eighteenth century, magnificent cultivated apple trees flourished along Michigan's shorelines, and the fruit was savored fresh and as cider and brandy.

Ribbon Farms

Antoine de la Mothe Cadillac established Fort Pontchartrain du Détroit in 1701 as a New France beachhead at the "gateway to the west" to act as a military post and control the fur trade. In a letter to the king of France dated September 25, 1702, he wrote of the fertile landscape along the strait, a narrow waterway that the French called *le détroit*:

> On the banks and round about the clusters of timber there is an infinite number of fruit trees, chiefly plums and apples. They are so well laid out that they might be taken for orchards planted by the hand of a gardener. The apples are of medium size, too acid...All the fruit trees in general are loaded with their fruit; there is reason to believe if these trees were grafted, pruned and well cultivated, their fruit would be much better, and that it might be made good fruit.

Cadillac planted a garden and grapevines and grew wheat, corn and peas. Native Americans were invited to settle by the fort and encouraged to plant extensively. Within a few years, land east of the fort was granted to newly arrived colonists to establish farms apportioned as two to four arpents wide on the river and twenty arpents straight back into wilderness. (One French arpent equals .8448 of an acre.) The farms fronted the strait for irrigation and for travel by canoe and bateaux in warmer months. In later years, many of the ribbon farms stretched back to eighty arpents.

Once granted a farm, landholders diligently cleared trees, tilled the soil, built log-hewn houses and planted seeds. During the early settlement years, Cadillac acquired "fruit trees in boxes," which were likely seedlings or grafted trees from Montréal or Québec, Canada. Unfortunately, Cadillac surely did not get to savor the harvest. In 1711, Cadillac, who had been removed from his Detroit post, left the area before the trees could bear fruit.

As the settlement expanded, the planting of apple, pear, peach and cherry orchards was a common practice inspired by "La Belle France," as was the way of life of tending orchards, eating fresh fruit and enjoying cider and brandy. Once trees bore fruit, fresh pears and the fruit as brandy, in particular, were great sources of French pride. The hardy apples, easily preserved in cool dugouts, were essential for balancing diets of wild game, fowl, whitefish and johnnycakes. Apples were sliced and dried in the heat of the sun; fire-roasted for sauce, pudding and apple butter; and crushed by large stones for cider, which naturally fermented. Hard cider lasted well into

Veritable Paradise of Fruits

the winter and was a staple even for young children as a slightly diluted drink called ciderkin. The fermented cider was also made into vinegar, over an aging process, and applejack, a brandy-like libation made by freezing cider outdoors. Exposed to winter's chill, the apple's essence naturally separated and, over time, evolved into a potent alcoholic drink. Cider was integral to the pioneers' lives and was sipped at every meal all year round. Cider and applejack quenched thirsts at a time when water quality was questionable.

Before long, two hundred inhabitants and a few cattle and horses lived on 350 acres of settled land. When the native Huron conceded land west of the Detroit settlement to the French in the 1740s, more ribbon farms were granted, and fruit trees from the "enchanted garden of Europe" were planted between Detroit and Monroe and across the Detroit River. In spring, it was reported, the fragrance of apple blossoms flooded the senses. Exquisitely flavored apple varieties that eventually grew included a deep red skin-to-core apple, which might have been the Bourassa apple; Pomme d'Api; Pomme Gris; Calville Blanc d'Hiver; Calville Rouge d'Automn; Detroit Red; and Fameuse, originally called Pomme de Neige. Though very few of the agrarian-minded French settlers owned titles to the lands they occupied, they remained devoted to the soil through the French and Indian War and the British occupancy of Detroit. Amid territory turmoil, frontier fever, ague and long winters, these resilient pioneers embraced this wild land and lived on a diet of venison, turkey, whitefish, wheat, corn, beans, squash and fruit.

As shared by Detroit historian Silas Farmer, a French memoir written in the mid-1700s remarked on the apple bounty of Grosse Ile: "…an extraordinary quantity of apple trees on this island, and those who have seen the apples on the ground say they are more than half a foot deep; the apple trees are planted as if methodically and the apples are as large as small pippins."

Meanwhile, inland of Lac des Illinois (later renamed Lake Michigan), missionary Pere Claude-Jean Allouez, who called the big lake "Lac St. Joseph," had established Jesuit Mission de Saint Joseph in 1686 near what is now known as Niles and the St. Joseph River. There, Allouez, followed soon by Father Claude Aveneau, worked alongside the indigenous Miami to farm the soil, sowing grain and vegetable seeds by the river to complement the native edible flora. A few years later, it was taken over as a French military and fur trading post called Fort St. Joseph. The St. Joseph River, named so in 1703, was a main thoroughfare for water travel between the Great Lakes and the Gulf of Mexico, connecting via the Mississippi River with a portage. By 1730, Monsieur Chevalier, a French lieutenant, was enjoying apples from his orchard to supplement the native diet that also included geese, duck, turtle, grapes, berries and sassafras, which was made into a drink. Chocolate, sugar and cheese from Montréal, Canada, and maple sugar, rice and cranberries from Fort Michilimackinac by the Straits of Mackinac were brought in by canoe.

The most significant growth, though, was in Detroit. In 1782, just before the end of the Revolutionary War, the settlement had 13,770 acres under cultivation and was undeniably a sight to see. In the 1790s, Isaac Weld Jr. noted:

> *The country abounds with peach, apple, and cherry orchards the richest I ever beheld; in many of them the trees, loaded with large apples of various dyes, appeared bent down in to the very water. They have many different forts of excellent apples...there is one far superior to all the rest, and which is held in great estimation, called the pomme caille. I do not recollect to have seen it in any other part of the world, though doubtless it is not peculiar to this neighborhood. It is of an extraordinary large size, and deep red colour; not confined merely to the skin, but extending to the very core of the apple; if the skin is taken off delicately, the fruit appears nearly as red as when entire.*

After the Jay Treaty finally eradicated Britain from the Northwest Territory in 1796, Detroit and its inhabitants became part of the United States of America. Brave adventurers from eastern colonies trekked for days and weeks to reach this exciting territory. Detroit's culture began to change, and many French families chose to relocate across the river in Canada. Those who stayed became integral to the character of Detroit of the 1800s, encompassing English, Irish, German, Polish, Scottish and Italian cultures. French-constructed cider mills gained a reputation for high-quality cider. Apples were crushed in massive wooden troughs by a six- to eight-foot diameter stone wheel that was pulled by a horse around a center axis. Apple pomace, which is solid remains, was fed to animals and used as garden compost, while the extracted juice was savored, stored for later enjoyment and exported from the harbor.

In the summer of 1803, under the directive of President Thomas Jefferson, Meriwether Lewis floated down the Ohio River to meet up with William Clark in Indiana to embark on an exploratory journey to the Pacific Ocean. Along the way, Lewis connected with John Armstrong of Cincinnati. Lewis asked Armstrong to send scions of fruit trees to Jefferson for his landscape at Monticello, which he had begun to develop in 1769. In February 1804, Armstrong sent Jefferson six scions individually wrapped with colorful strings, four of which were from Detroit's apple trees that had been sent to him two years prior. The fruit was described as "large white apple, large red apple, pumgray and calvit." According to Peter Hatch, former director of gardens and grounds for Thomas Jefferson's Monticello, the calvit was likely Calville Blanc d'Hiver, the pumgray was possibly Pomme Gris and the "large red apple" might have been Detroit Red, though there was never proof that it was Detroit's namesake apple. Armstrong described the "pumgray" as "much admired" and commented that it stored well year round. He described the "calvit" as "without comparison the best apple that ever was Eaten." The white apple might have been Yellow Bellflower.

Michigan Territory

After the turn of the century, Detroit's original farm homesteads and settlement were showing their age. When a fire swept Detroit in 1805, just a few weeks before the new Michigan Territory governor and judges arrived, it was an opportune time to redefine Detroit's architecture. Initially, this was fairly difficult as, according to Judge Augustus Woodward in an 1806 letter, there were just eight legal land titles in the territory. In another 1806 letter, Woodward indicated that there were 422 farms and settlements in the regions, 77 of which abutted the Detroit River. After much discussion and evidence of "squatters' rights" to settlements, land patents were sorted out, and new generations of families tilled the soil.

Wayne County had grown in population, and the new Catholepistemiad, or University of Michigania, was established in 1817. Overall, though, the Michigan Territory progressed slowly due to the difficulty of land travel. It took roughly thirty days to travel from the East Coast to Detroit. It didn't help when, in 1816, Edward Tiffin, surveyor general of the Northwest Territory, reported to the General Land Office of the United States that Michigan consisted largely of swamps, lakes and poor, sandy soil—not exactly enticing to adventurers desiring to settle the Wild West. Those who resided here knew the soils were fertile and the waterways flush with fish. They carved out a living, relying on the exchange of goods and cider and fish exports.

Michigan Territory remained largely rural and unpopulated until the opening of the Erie Canal in 1825 and the arrival of steamship transportation. That same year, surveying began to develop the Chicago Road (now U.S. 12), which was the Old Sauk Trail that had been historically used by mastodons, Native Americans and seigneurs. Adventurers caught the "Michigan Fever" and traveled here by the thousands by ship, on foot and via wagon, intent on reaching the Michigan Territory.

William Woodbridge, secretary of the Michigan Territory, planted two thousand apple trees in 1825 on his Detroit farm. He grafted rootstock with Detroit's heritage varietals—Fameuse (Snow), Calville Blanc d'Hiver and Calville Rouge d'Automne ("White and Red Calville"), Detroit Red and Pomme Gris—and planted New England varietals, including Rhode Island Greening, Baldwin, Esopus Spitzenburg, Roxbury Russet and Twenty Ounce.

By now, John Chapman's seedling trees had firmly rooted throughout the Ohio Valley. Chapman, better known as Johnny Appleseed, had walked by foot from the East Coast and sown many apple seeds collected from cider mills along his journey. It is possible that when Indiana, Ohio or Pennsylvania settlers migrated to Michigan in the mid- to late 1800s, they brought seedlings from some of Chapman's trees, though the bitter apples the trees produced would have been used for hard cider or applejack. Michigan's first apple orchards are firmly attributable to its earliest French settlers.

In 1830, the population was nearly 32,000. Within the decade, more than 212,000 people had settled in the territory. Entrepreneurs established nurseries to supply fruit trees to the growing population of homesteaders. In 1840, "Variety" White of Monroe established a nursery with grafted fruit trees from the East Coast. Another early nursery was established in Ypsilanti with 25,000 trees of 130 apple varieties from New York. That nursery supplied trees to farmsteads in the southwest and southeast counties and as far north as Kent and Genesee Counties. Additional roadway routes were established, and finally in 1873, gravel was implemented to improve travel. This aided settlers in trade and in transporting fruit to markets.

Lake Michigan Fruit Belt

Beyond the Detroit region, enthusiastic land seekers were granted patents to settle west Michigan, where old orchards and native edibles, like the pawpaw fruit, thrived along rivers and in oak openings. More than 250,000 acres of land were sold at $1.25 per acre between 1831 and 1834 from the White Pigeon Land Office in southwest Michigan.

Veritable Paradise of Fruits

To no scenery of our country that I have yet seen is the term "Arcadia" more applicable than to the rich and fairy landscape on the west side of the peninsula, watered by the Kalamazoo and the St. Joseph.
—*nineteenth-century travel writer*

Long before this flood of new settlers, William Burnett, who had settled in the wilderness along the St. Joseph River around 1780, had a thriving orchard and trade business with the Native Americans. Well into the nineteenth century, Burnett's apple orchard thrived and was admired by those who settled after him. In 1839, Benjamin Hoyt, a southwest banker and nurseryman, grew peaches and apples in the same region. Inspired by the sight of high-quality orchard fruit growing in the region, he sent a load of peaches by steamship to Chicago and launched Michigan into the commercial fruit industry of the nineteenth century.

Homesteaders began to settle along the Grand River in Kent County in the 1830s, planting apple seeds and grafting scions from the old orchards of Detroit on native rootstock.

Hard Cider on Tap

Prospering in rural Michigan meant following the rhythms of the seasons. Pioneers did sugaring in the spring and cidering in the fall and relied on heart and grit to survive in this undeveloped land. Most arrived with just an axe, few clothes and a satchel of seeds and settled in deeply forested land rampant with wildlife. Optimists brought scions from their favorite apple trees.

Once the trees produced fruit, autumn frontier fun included gatherings to pare down apples and socialize in preparation of winter. Apples were sliced, dried and strung up; sauced and stored in crocks in cupboards; and crushed into cider that fermented and kept during the frigid months.

Three years after the Michigan Territory joined the union in 1837, residents voted for the first time in the national presidential election. When the 1840 campaign rolled around, the voting men were boisterous in their support of their hero and "every-man" candidate, General William Harrison. Their enthusiasm helped launch the historic "Log Cabin and Hard Cider Campaign."

Historian Friend Palmer, who wrote of the early days of Detroit, helped cut logs to build a cabin for the Whigs of Detroit in 1840. Though Palmer wasn't old enough to vote, he recalled the novelty of the log cabin built on a vacant lot in Detroit as a mass meeting place to rally support for Harrison's campaign. Oxen hauled logs into the city, which were then stacked to make a primitive cabin that was "decorated with dried coonskins…and festoons of dried apples, dried pumpkins, and corn ears…while a barrel of hard cider, on tap, occupied a prominent position." Other log cabin Whig gatherings took place in Lower Michigan; though, perhaps, Detroit's was most notable for the procession of log cabins on wheels and the dispensing of hard cider in tin cups and gourds to the rambunctious crowd. The campaign tactic was a success in Michigan and beyond. "Tippecanoe and Tyler Too" took office while promoting that "the highest measure of happiness could be found in a log cabin with an abundant supply of hard cider."

Michigan Blooms

In Saginaw Valley, century-old apple trees in the region prompted settlers to plant new trees in 1833. Besides cultivated fruit, landowners raised dairy cows, hogs and vegetables. In Kent County, Abel Page and his son purchased ten thousand root grafts from Monroe, primarily focused on growing apple trees. Their nursery operated for twenty years and, at its pinnacle, had 250,000 trees. In 1849, Michigan held its first state fair in Detroit, where agricultural products and new inventions were unveiled to the public, with premiums awarded for the finest efforts. The following year, the Michigan Agricultural Society recognized farmers with "premiums" for their fine orchards and vineyards. One such farmer was Linus Cone of Avon, near Troy. He was well regarded as a writer and orchardist and, particularly, for his well-pruned apple trees that generated fifty bushels of apples annually for his family. In 1824, seventeen-year-old Cone had walked alone from New York to Oakland County, eventually owning 160 acres of land.

As land was cleared for timber, the commercial fruit industry blossomed from Kalamazoo and St. Joseph to the Straits of Mackinac. By the 1840s, fruit trees grown as seedlings and from grafted rootstock were in every county in Lower Michigan. Seeds of Fameuse, one of Detroit's first cultivated apple varieties, were often tucked in travelers' sacks before trekking to claim new land in feral regions. One of these seeds grew into a beautiful-tasting, aromatic apple. Resembling Fameuse, the apple closely matched its parent in looks and slightly tart taste but offered a bit more spice and hardiness. Discovered by Beebe Truesdell in 1850, the apple became known as Shiawassee Beauty. It was much admired and described as "vigorous and hearty," with "nature's own flavoring stored in the fruit." This tree was propagated, and the apple variety can be found in some orchards today, including on North Manitou Island in Lake Michigan off the shores of Leland. It is also cultivated at farms specializing in heritage varieties.

In 1858, near South Haven, Liberty Hyde Bailey Jr. was born on a farm amid a woodsy setting. Over time, he became known for his skills at apple tree grafting. Nature's wild beauty inspired Bailey, who wrote poetry, studied horticulture at Michigan State College and became one of America's most prolific horticulturalists and literary writers. Michigan's landscape dramatically evolved in the thirty years Bailey lived in Michigan before accepting a horticultural post at Cornell University.

Northward

Pressured by the U.S. government, Native Americans relinquished claim to land north of the Grand River. Soon after, European immigrants arrived en masse and established homesteads along waterways. In the late 1830s to mid-1850s, settlers, many of German descent, inhabited a land ridge north of Grand Rapids and actively timbered the region. This region would eventually be known as the Fruit Ridge.

It would later be discovered that apple trees thrived in the naturally moist clay loam of this particular 158-square-mile region. The trees flourished on the rolling hills and benefited from a temperate climate eight hundred feet above Lake Michigan. Remarkably, roughly 65 to 70 percent of Michigan apples are grown on this ridge today. Several families who settled in the 1800s and early 1900s are still farming in that region. Families named Alt, Anderson, Armock, Baehre, Brechting, Brown, Dietrich, Dunneback, Ebers, Hill, Klein, Klenk, Kober, Kraft, Nyblad, Rasch, Reister, Schaefer, Schwallier, Steffens, Succop, Thome, Umlor and Youngquist, among others, are still tilling the soil or working in the fruit business.

"They followed the Grand River, looking for land similar to back home," said Harold Thome, a fourth-generation farmer, whose great-grandfather Michael Thoma settled in Alpine Township in 1846 after traveling through

An 1876 lithograph of Waterman farm (now Alt farm) taken from *Illustrated Historical Atlas of the Counties of Ottawa and Kent, Michigan*. A reproduction was made possible by the Grand Rapids Museum Association. *Courtesy of Alpine Township Historical Commission.*

Detroit and Westphalia, Michigan. From Trier, Germany, the Thoma family homesteaded 160 acres in Alpine Township and was listed on the original plat map for the community. The name evolved to Thome, and today, Harold's son Steve and grandson Mitch farm the same land.

Tools were few in the early days, and oxen were relied on heavily to clear land for the first orchards of peach, apple and plum trees. Barns were built by hand—molding felled lumber with the axe, adze and chisel. Peach orchards dominated the landscape, though it was also common for farmers to plant young apple trees amid the peach trees in an alternating diamond shape. When peach trees died after their average ten-year life span, apple trees matured and branched outward, filling the space. Before long, steam-powered cider mills were built along waterways.

In 1850, there were 36 farms growing agriculture on roughly 1,620 acres, much of it harvested and hauled to city markets. Within thirty years, there were 246 farms cultivating nearly 14,000 acres.

West of the ridge in Spring Lake, Hezekiah Smith, who had established a farm in 1849, gained a reputation as an excellent peach, apple and cucumber grower. A former slave, Smith found haven as a free man in the pretty lake community. Smith might appreciate that apples saved a man who had hid in the bottom of a large crate of apples in a safe home along the Underground Railroad near Kalamazoo. "Slave hunters" searched the home but couldn't find him. On their way out, they stopped to admire the apples, and the house owner offered a few for the road.

VERITABLE PARADISE OF FRUITS

An apple picker loads up his harvest bag at the Kent Brown farm in Alpine Township. *Courtesy of the Kent Brown family and Alpine Township Historical Commission.*

Wherever settlers went, fruit trees were planted. In 1865, similar to other counties, Benzie County established an agricultural society to aid settlers in their fruit pursuits. Within twenty years, regional growers were touting many apple varieties and showcasing an impressive array of apples at the annual meeting of state growers.

The region of Grand Traverse began to sprout in the late 1840s with the arrival of settlers wishing to live in the "delightful summer climate" and enjoy the "pure water" of the region. These land seekers discovered feral orchards dotting the landscape and shores, the old farms and gardens of the missionaries and the Native Americans who had grown fruit to supplement the diet of wild game and native vegetation. In 1869, apple trees, some of which were thirty feet tall, were discovered as far north as the Straights of Mackinac and Cheboygan and spotted on Mackinac Island. On Old Mission and Leelanau Peninsulas, one-hundred-year-old apple trees grown from seed were discovered, and many produced palatable apples. Early on,

MICHIGAN APPLES

A 1914 apple harvest in Wesley Smeltzer's orchards near Putney Corners. *Courtesy of Benzie Area Historical Museum.*

agriculturalists remarked on the beauty of the region and its prolific ability to grow a variety of fruit. And soon after the area was civilized, the bluffs were blanketed with orchards and vineyards.

The peninsulas were favored destinations of those seeking orchard lands. Reverend George Smith of Northport gained publicity in the *Grand Traverse Herald* for his high-quality apples. John Pulcipher, who settled in 1855 in Acme Township, was considered one of the "best farmers in the county" with one of the "finest and best improved farms."

In 1861, Judge Jonathan Gannett Ramsdell planted orchards and vineyards on the west arm of Grand Traverse Bay and grew peaches, grapes and apples, including Golden Russet, Rhode Island Greening, Wagener, Rambo and Red Astrachan. Judge Ramsdell later served as president of the Grand Traverse Union Agricultural Society and Michigan State Pomological Society.

Renowned fruit culturist George Parmalee, who had excelled at peach growing in the South Haven area, relocated to Grand Traverse County and planted more than one hundred acres of apple trees on his four-hundred-acre northern orchard.

In 1861, W. Golden, H.R. Haight and E.P. Ladd planted orchards of Golden Russet, Rhode Island Greening, Baldwin and Northern Spy on Old Mission Peninsula for the commercial market.

On neighboring South and North Manitou Islands, enterprising pioneers cleared land to supply timber to passing steamers for fuel. In 1881, war

VERITABLE PARADISE OF FRUITS

An 1894 apple harvest at W. Golden orchard on Old Mission Peninsula. *Courtesy of Archives of Michigan.*

veteran Frederic Beuham homesteaded 160 acres on North Manitou Island. He initially planted 500 fruit trees and grapevines. He partnered with William Stark of Stark Bro's Nurseries and Orchards of Missouri in 1894 and expanded his acreage to grow 2,500 apple trees and 1,500 pear trees. Within five years, Beuham had lost control of his land to Stark, who then sold it to the Newhalls. The orchard, now within Sleeping Bear Dunes National Lakeshore, bears apples on roughly 700 to 1,000 apple trees, some of which grow Ben Davis, Grimes Golden, Jonathan, Shiawassee Beauty, Stark Delicious (original strain of Red Delicious), Wagener and Wealthy apples, all American heritage varieties.

Around the state, families established farming traditions that would last generations. These include, between the early 1800s to roughly 1920, Westview Orchards and Adventure Farm (Washington Township), Keeney Orchard (Tipton), Wiard's Orchards and Country Fair (Ypsilanti), Phillips Orchards (St. Johns), Overhiser Orchards (South Haven), Smeltzer Orchards (Frankfort), Johnson Orchards (Old Mission Peninsula), Almar Orchards (Flushing), Wittenbach Orchards (Belding), Klackle Orchards (Belding), Moelker Orchards Farm Market and Bakery (Grand Rapids), Robinette's Apple Haus and Winery (Grand Rapids), Wells Orchards (Grand Rapids) and Uncle John's Cider Mill and Fruit House Winery (St. Johns).

Fruit to Markets

Enterprising Michiganders utilized the evolving modes of transportation to ship fruit to markets in Chicago and Milwaukee. Rail routes sprouted from city hubs and eventually connected from coast to coast. In 1868, a patent for an improved "icebox on wheels," a refrigerated rail car for shipping fruit, fish and meat, was developed by fish dealer William Davis of Detroit. Apple barrels by the thousands were shipped into the Chicago market in the late 1870s. Apples traveled fairly well, and soon farmers began to better appreciate the apple's profit potential. Watervliet farmer Harvey Sherwood converted timber-cleared land to orchards and earned the nickname "Apple King" for his significant harvests and shipments into Chicago. Two years of apple shipments from Fennville tallied nine thousand apple barrels, and thousands of barrels shipped from Ionia. Many communities, like Chelsea, Saline and Ypsilanti, shipped apples to bustling city markets. In 1889, 13.16 million bushels of apples were harvested, compared to 216,311 bushels of peaches.

While the refrigerated car was a boon to the apple industry, Michigan's peach industry was adversely affected. Georgia farmers with their longer growing season could send peaches into the mighty Chicago market faster and more often than Michigan growers, essentially winning over this highly populated metropolitan city with their quality crop.

As early as 1802, fruit and vegetables had been sold or traded at market. In 1816, the Woodward Avenue Market opened and stayed in business until 1835. Six years later, the Detroit Farmers Market opened. It eventually moved to its present location at Eastern Market in 1891. Across the state, the Benton Harbor Market opened in 1860, rapidly becoming the world's largest cash-to-grower market. Both locations were advantageous for steamship and rail shipments of fruit throughout the United States. In 1865, both sides of the St. Joseph River through the Benton Harbor and St. Joseph region grew 70,000 apple trees, more than 200,000 peach trees, 10,000 cherry trees and quince and plum trees, as well as grapevines and strawberries. Farmers carted apples to the markets of Detroit, Benton Harbor and Grand Rapids to sell directly to grocers and to ship to Chicago, Milwaukee and Boston to fulfill the insatiable demand for Michigan's agricultural bounty.

As cities grew, constructed roads were required to serve the growing population. In 1857, legislation passed to enlarge the corporate limits of Detroit. This growth adversely affected William Woodbridge's impressive farm, which combined four of the old French ribbon farms of the 1700s and was further nurtured by Woodbridge's son. Woodbridge fought to keep what remained of his farm intact, arguing that "a substantial wealth of state consists in its agriculture." He lost the battle, and a new travel route through his farm destroyed many grafted fruit trees. Years earlier, he had agreed to a railroad right-of-way across the front of his property. Before his death, he remarked that "the passion for creating city lots would eventually destroy all the farms in the vicinity of Detroit."

Due to the evolution of transportation, the call for settlers rallied on. Michigan was a destination with opportunities in agriculture, mining, logging and industry. With established water, rail and road routes, travel from New York to Michigan took as little as two days. A century ago, it had taken a month of hardships. Approximately 700,000 immigrants arrived between 1860 and 1900, more than half of them foreign-born. Settlements included the new capital of Lansing.

Cheapest Lands in the World, considering the quick cultivation, varied production of high quality, and practically no time from the great markets of Chicago, Milwaukee, Detroit, etc. are in the famous fruit belt of Michigan along the eastern shore of Lake Michigan. A populated region with schools, churches, railroads, steamboat lines, telegraphs. Millions of people to buy all fruit, vegetables, garden truck as fast as it grows, and transportation ready, quick and cheap enough to get it to them. $5 to $20 per acre.
—American Gardening, *1894*

In 1914, the Cohodas Brothers Fruit Company was established in Houghton. Nineteen-year-old Sam Cohodas and his brother Harry would soon run their father's wholesale fruit business and grow it to become the third-largest produce company in the country. Eventually, Sam owned apple orchards to supply his fruit business. One orchard was on land north of Manistee that had been timbered by William Douglas. The farm was on an elevated ridge of up to nine hundred feet that benefited from tempering Lake Michigan breezes. Thanks to Douglas's progressive timber business, the farm was reached by narrow-gauge rail. It became part of the Manistee and Northeastern Railroad, which eventually expedited Cohodas's fruit to metropolitan areas.

Sam Cohodas cultivated his successes into other ventures, acquiring orchards in Washington and California as well. His apples supplied the fruit for Apple Keg, an apple juice that he distributed throughout the nation and to American troops serving overseas during the Second World War. Cohodas's canning operation in Elberta in northwest Michigan processed the juice, which was "a blend of three and, sometimes, four varieties" that was "twice filtered and twice

MICHIGAN APPLES

Cohodas Brothers Fruit Co. produced popular Apple Keg juice at Elberta Packing Co. *Courtesy of Central Upper Peninsula and Northern Michigan University Archives.*

pasteurized" for a quality product. Cohodas served on many organizations throughout his lifetime, including as director of the International Apple Association. In the early 1980s, Cohodas visited his old Manistee farm, renamed Lakeview, and chatted with Dennis Mackey, who managed the orchards at the time. Mackey recalled Cohodas's delight in seeing the trees. He would surely appreciate that the farm operates today as Douglas Valley and that a block of untamed Northern Spy trees from Cohodas's era still thrive.

THE STUDY OF AGRICULTURE

As the farm focus changed from growing food for just the family to providing food for the mass market, orchardists were compelled to gather together to share experiences maintaining older orchards and establishing new ones.

The study of agriculture was crucial to the growth of Michigan's fruit industry. Crop challenges included apple scab and the codling moth, as well as extreme situations and temperatures. In 1834, the seventeen-year locusts destroyed young orchards in Lenawee County. In 1871, the same year of the great Chicago Fire, fires ignited homes and orchards throughout Michigan due to the drought. On the plus side for settlers, more land became available for cultivation due to the felling of timber to rebuild cities of Michigan and buildings in Chicago. In 1872 and 1897, temperature swings and extreme frosts crushed the crop load down to fewer than four million bushels.

Another rising concern was the discovery that many apples had many names. It was essential to streamline nomenclatures to successfully market an apple variety to the consumer. Fortunately, horticulturist Theodatus Timothy Lyon had been focused on identifying apple varieties since opening a nursery in Plymouth in 1844. As a youth living in New York, Lyon had experimented with grafting, a passion ignited after inserting cuttings from his favorite apple trees into others. With the help of expert nurserymen, Lyon worked diligently to identify Michigan varieties.

Michigan's rapidly expanding fruit industry was aided by the opening of the Agricultural College of the State of Michigan in 1855 (Michigan State University), which later became the nation's first land grant institution. It required students to work on farms for agricultural research experience.

In 1870, the Kent County Agricultural Society hosted its first fair in Grand Rapids, and 150 apple varieties of all sizes, shapes, color and quality were on display. Approximately twenty thousand people attended the exhibit.

The old settlers who have borne the toil and broke the ground, were thankful to see, this day, the fruition of their hopes. Here was the result of their labor, and here was the pledge of the bounty for the future.
—*1870 Michigan State Board of Agriculture*

Also that year, the Michigan State Pomological Society, an organization of fruit farmers and specialists, was initiated, and apples were displayed at the first April meeting. It was reported that the "large, bright-looking" Baldwins, "genuine" Roxbury Russets, "brotherly-looking" Jonathans and "splendid" Wageners were "attractive to the eye and delicious to taste."

The soil, climate, and geographical position of the State of Michigan have shown that she is a favored region, and well adapted to the cultivation and growth of all fruits suitable to a Northern and temperate clime...to-day, as we look over a vast country, from lake to lake, we see large and flourishing orchards of apple, pear, peach and cherry...it is desirable that those engaged in fruit culture should seek a closer connection with each other, and should establish an organization which should directly represent their common interests.
—*1872 Michigan State Pomological Society*

In the first report issued by the Michigan State Pomological Society in 1872, farms and specific cultivars of fruit were noted for fairing well. A few notables include the nursery of Noah P. Husted of Lowell, who, in 1862 with James D. Husted, had planted forty thousand apple trees and developed test orchards and vineyards with a strong focus on grafting scions onto rootstock to determine what varieties would grow most successfully here.

Husted was excited about the Wagener, "considered a marvel of hardihood" with good cooking qualities like the Rhode Island Greening. Although he did not favor the Esopus Spitzenburg, other fruit growers considered it a "very fine fruit," and most praised the Russian apples, Red Astrachan and Duchess of Oldenburg. Husted was also praised for his "esteemed" Baldwin, Northern Spy, Hubbardston Nonesuch, Tolman Sweet and Golden Russet. Other favorites of growers included the "splendid" Swaar, "best autumn" Porter and the excellent Jonathan and Snow (Fameuse). At this meeting, the men discussed how best to store apples, with one suggesting, "Heap them up to sweat, then pack them in buckwheat, chaff or bran."

Also remarked on was that Michigan was the Fruit Belt and that the state's apples were being shipped to Liverpool, England, and "recognized as the very best, and sought after in the principal markets of the United States, retained on the streets of New York City, and placarded on busy corners of Denver."

At the 1873 pomological meeting, "the apple, in all its glory, was king of the occasion," with all parts of the state represented from "Grand Traverse to Berrien, Monroe to Saginaw."

Seeds of Ingenuity

As cultured and educated settlers flooded Michigan to claim land, industry followed with the opening of cider mills, vinegar factories, farming equipment suppliers and fruit basket makers.

In 1870, the first Alden evaporator was invented in New York. This "apple drier" had a significant impact on the industry, and fruit farmers were quick to adopt the technology, which stopped fermentation and maintained the fresh taste of the apple. An Alden factory opened in Spring Lake and several more in the Grand Rapids region. In 1889, eighteen fruit evaporators operated in the region, and $55,000 in dried fruit was shipped.

Meanwhile, apple cider vinegar factories opened, including Wiard and Son of Ypsilanti. Vinegar was produced for four generations and was last managed by Phil "Pete" Wiard until the 1990s, when cost-cutting competitors made business difficult. At their peak, the Wiards made 200,000 gallons of vinegar a year on their farm, processed another 500,000 gallons into bottles and distributed the vinegar for twenty-six private labels. Since then, the vinegar tanks and pumps have been repurposed into Wiard's haunted theme barns.

Transcendent Apple

In the late 1800s, Niles citizen George Hoppin farmed his father's land near the St. Joseph River and an "old fort and mission," where he found three-foot-diameter apple trees more than a century old still bearing fruit. These ancient trees were a tribute to the region's early tenants of Fort St. Joseph.

> *There are pomological traces still surviving in our borders of the French missionary pioneers of our state. We are told that when Father Marquette and his Jesuit brothers paddled around our beautiful peninsulas, over two hundred years ago, dropping here and there the seeds of a few apple or pear trees along with the "seed of the Word," they little dreamed of the great future that was indicated for Michigan, in the thrifty growth of seedling fruit trees that were to spring up along their pathway. A few of these old trees and some that were planted a few years later, are still found in various places of the shores of Erie, Huron and Lake Michigan. They are healthy and strong, bearing crops of fruit, monuments of the adaptability of our State to the cultivation of the apple and pear.*
> *—Senator Thomas W. Ferry in regard to the growth and progress of Michigan to the Pioneer Society of the State of Michigan in 1882*

Apple trees adapted well to Michigan's cold climate. In 1889, it was recorded that apple bushels exceeded thirteen million. In 1893, Chicago hosted the World's Columbian Exposition, and Michigan was proudly represented by its fruit culturists.

> *Regal, indeed, have been the gifts of nature to our State. She has a rich, warm and mellow soil, which bountifully rewards the toil of the husbandman, and yearly fills to overflowing his granaries and barns. She has a climate so propitious that a large part of her territory is a veritable paradise of fruits, when Heaven kindly draws the sting of frost from the keen west wind, so that the breezes soft*

Jim Robinette's relatives pick apples in west Michigan in 1900. *Courtesy of Robinette's Apple Haus and Winery.*

as those of Eden woo the peach and the grape and the plum and the pear and the apple, and coax them to rejoice as with the autumnal splendor of their fruitage, which rivals that in the fabled Garden of the Hesperides.
—University of Michigan president James Angell dedicating the Michigan exhibit

It is the fragrant blossom of the native crab apple (*Pyrus coronaria*) that was deemed a most worthy symbol as the state flower. In a joint resolution in 1897, it became official. Just a year prior, the farmers had celebrated a bumper crop of apples and nearly twenty-three million bushels. Unfortunately, the next year brought a yield of fewer than four million bushels due to extreme frost. Not deterred by erratic Mother Nature, the agrarian-minded pioneers continued to plant orchards.

In 1900, 60 percent of Michigan's nearly 2.5 million residents lived in the country, and apple orchards manifested in every county. Combined, Allegan, Berrien, Van Buren and Kent Counties grew nearly four million bushels, a quarter of all apples in the state. The apple crop equaled the value of the peach crop, with the largest apple orchard growing fourteen thousand trees in Berrien County. In 1906, extreme fall temperatures killed 90 percent of the peach trees and damaged the remainder. This phenomenon further raised the status of the apple.

RECIPE

Apple Butter
Courtesy of King Orchards of Central Lake

10 pounds of tart and/or cooking apples (Lodi, Cortland, Idared, McIntosh, Northern Spy, Rome)
4–6 cups apple cider
3 cups sugar
2 tablespoons ground cinnamon
1 teaspoon cloves
½ teaspoon allspice

Quarter and core apples. Do not peel. Place in a large cooking pot and add cider to cover the apples. Cover and cook over medium heat until tender, about 20 minutes. Use a food mill or push through a sieve. Apple pulp should measure about 3 quarts. Stir remaining ingredients into pulp. Pour into roasting pan. Bake uncovered at 275 degrees for about 6 to 8 hours. Stir occasionally until it is a rich, dark brown. It should be thick when scooped. Spoon into jars. Seal and refrigerate.

Chapter 2
MARKETS BLOOM

After decades of purposeful fruit growing, farmers of the twentieth century needed advice on how to deal with fruit infestations, aging orchards, fickle weather and market demand for high-quality fruit. Farmers consulted state horticultural and pomological experts, especially those connected with the Michigan Agricultural College Extension, which had been established in the last few years of the previous century.

In 1909, a multi-year study by the agricultural college extension addressed the question "Can the general farmer afford to grow apples?" It identified tips for improving orchards and evaluated the costs of management. The study included the evaluation of thirty- to sixty-year-old orchards growing varieties like Stark Delicious, Maiden Blush, Winter Rambo, Ben Davis, Steele Red, Baldwin, Golden Russet, Rhode Island Greening, Winter Pippin, Fameuse (Snow), Yellow Transparent, Northern Spy, Jonathan, Fall Pippin, King, Bellflower, Tolman Sweet and Red Astrachan.

The study identified the need for protecting orchards against such issues as the San Jose scale pest by applying "Bordeaux" spray mixture, a French invention of the previous century that combines water, lime and copper sulfate. Selling apples commercially was a different focus than growing apples for family consumption. Apples needed to be uniform in size, attractive, free of worms and blemishes and able to keep well in storage barrels. Farmers began implementing the research station's tested strategies to improve fruit for the commercial market.

In 1914, apple and cherry farmers felt short-changed on their fruit sales, finding it more and more difficult to compete in other regions. Farmers realized that to make a profit, they needed to know how to market their fruit. Already, New York and other states had adopted laws establishing regulations for "grades" of fruit based on condition and uniformity. To compete beyond Michigan, the Grand Traverse Fruit and Produce Exchange adopted New York's apple-packing laws and hired a New Yorker to get Michigan apples into other markets, including European "car-lot" markets. Five years later, the Michigan Farm Bureau was established to provide support to farmers to address a range of challenges.

Michigan was the fastest-growing state between 1910 and 1930, in large part due to the land opportunities and fledgling automotive industry. Immigrants arrived in droves to Michigan, responding to ads promoting "wealth in Michigan farms" by the Pere Marquette Railroad and "one million acres of lands in the state of Michigan for sale." With landownership possibilities and industry work in big cities, including at Henry Ford's assembly line plant, Michigan was more attractive than ever.

The Robinettes carted apples to the Fulton Street market in downtown Grand Rapids in 1912. *Courtesy of Robinette's Apple Haus and Winery.*

Markets Bloom

Barzilla Robinette, who had bought farmland in 1911, quickly established wholesale accounts in downtown Grand Rapids. After harvests, he and son Edward carted apples by wagon to the Fulton Street market. Oftentimes, they would sell their produce before the market opened, meeting fruit vendors along the way.

Innovations

As Ford's automotive assembly line revved up, the commercial fruit industry welcomed new innovations. In 1911, Henry Kraft sold ninety-two bushels of his Baldwin apples to a Grand Rapids market for ninety-five cents a bushel. The following spring, he learned that the buyer had sorted his apples into two grades and sold the top grade for three dollars a bushel and the second grade for two dollars a bushel. This prompted him to build a storage unit that was insulated for winter protection and featured large, wide doors that could be opened to keep the apples cool. Storing apples and selling them direct lent to better profits. The strategy worked, and soon after Kraft and other area farmers built additional storage units. Kraft's innovation shifted the commercial apple industry in an entirely fresh direction and was the predecessor to future storage innovations to prolong the apple's freshness beyond the fall harvest.

Simultaneously, a new opportunity arose to supply fruit to canning companies. The initial concept of preserving fruits and vegetables in containers had been developed and patented in Europe. It was after 1896 when canning was approached more scientifically—and the issue of botulism addressed—that it became a viable industry in Michigan. And in 1922, the Canners of Michigan promoted "the purity and wholesomeness of canned foods."

The polar regions and the tropics alike rejoice in the bounty of the temperate zone. The ship destined for a long voyage no more need fear the inroad of disease which often made a hell of the very stronghold of commerce. For all the hoarded gifts of summer live in the can, and the sunlight of August is preserved to make happy the frosts of December.
—F.W. Schultz, "Retrospect,"
in A History of the Canning Industry, *1914*

Jack Brown poses with apple bushels on his family's Alpine Township farm in the early 1900s. *Courtesy of the Kent Brown family and Alpine Township Historical Commission.*

Also notable was the onset of refrigeration that began to arrive in cities and suburbs in the late 1920s. By the early 1930s, several Ridge farmers had converted their storage units to mechanical refrigeration.

In 1921, the Baldwin apple variety accounted for a quarter of the apple crop in west Michigan, with Northern Spy and Duchess of Oldenburg accounting for an additional 14 percent each of the crop. Ben Davis, Wagener, Rhode Island Greening, Wealthy and Jonathan were also important crop apples, with Ben Davis predominantly grown in older orchards.

Markets Bloom

In 1927, a copious crop of apples prompted Jack and Aleta Brown to launch a wholesale fruit business in Grand Rapids. Initially, the Browns sold apples from their garage direct to grocery stores. They traveled from farm to farm collecting, sorting, loading and filling baskets into their truck for distribution from their "market," and before long, their fledgling business expanded. Jack fell ill and did not continue to manage the fast-growing business. Aleta remained in control of the operation.

"Mrs. Brown was probably the first female fruit marketer. She was a fearless businesswoman and could make growers shake in their boots," said Julia Baehre Rothwell, who grew up in Kent City within the Fruit Ridge. "She controlled the destiny of many Ridge growers and was a legend and an incredibly successful businesswoman." Now a grower cooperative, Jack Brown Produce is one of Michigan's largest fresh-to-market sales and packing businesses.

"Mrs. Brown was an absolute legend in the industry," agreed current president John Schaefer. "She was successful in a man's world, which is what the produce industry was at the time. We were all a bunch of hard-headed, stubborn Germans who operated businesses and thought we could dictate to Aleta. That was not the case. She was able to get the grower board and stockholders to march together for the good of everybody. It took someone of her special talents to be able to do that. She had a pretty good left hook, too. More than one grower who ran afoul of the rules was laid out on the office floor."

At the same time that the Browns started their fruit business, the Fremont Canning Company, which opened in 1901 to can the region's agriculture, took an innovative turn in 1927 when Dorothy Gerber started straining peas for her daughter. The laborious task prompted her husband to use their canning company's process to strain vegetables and fruits. Employees were keen on the idea and began requesting food for their babies, prompting the launch of Gerber Baby Foods in 1928. Ever since, Gerber (now a brand of Nestlé) has partnered with west Michigan farms for freshly harvested apples for processing into shelf-stable baby food.

In 1929, fruit growers on the land ridge north of Grand Rapids collaborated to market their region. They called themselves the Peach Ridge Fruit Growers Association, consisting of farms "west of Grand Rapids, bordered on the north by Bailey Road, the south by Leonard Street and a line passing through Coopersville and Ravenna on the west." This region would later be known as the "Fruit Ridge" or "Ridge."

Prohibition

Henry Ford's requirement that workers at his plant uphold the strict moral code of refraining from alcohol consumption played a role in Michigan's active temperance activities. By the start of national Prohibition in 1919, Michigan had already undergone a period of prohibition in the mid-1800s and was the first state to go "dry" a second time when the Eighteenth Amendment was ratified. Thanks to the language of the Volstead Act, limited hard cider production was legal, as fruit could be preserved by fermentation. And though the "wets" prevailed during this failed national thirteen-year experiment, the "drys" fought hard, many of whom axed down cider trees in an effort to end hard cider and spirit consumption. Hidden orchards on family farms were commonplace, though, and apples and other fruit were made into plenty of batches of hooch.

By the 1920s, D.H. Day had become a formidable businessman in Glen Haven. He owned the village and thousands of cherry trees that warranted a canning company to process and ship fruit to busy city markets by steamships. To maintain his well-regarded reputation, Day tucked apple trees within hidden pockets of his land to supply fruit for moonshine for his employees and himself. The fall harvest supplied the libation during the thirteen years of Prohibition. Upon his death, Day left a remarkable legacy and donated land for conservation, which became Michigan's first state park, D.H. Day Campground in the Sleeping Bear Dunes National Lakeshore. Amid this breathtaking landscape just a tad southeast of Glen Haven stands the last of Day's Prohibition-era apple trees—a knobby, twisting tree with a wild canopy of fall apples.

Even though many Michiganders found a way to indulge in alcohol during Prohibition, nationally, Coca-Cola sales tripled, Welch's grape juice generated record sales and root beer became popular as a "wholesome temperance drink." In April 1933, Michigan was first in the nation to repeal Prohibition—though, ironically, with the passing of the Twenty-first Amendment, it was often harder to get alcohol. The new amendment gave control of alcohol regulation to the states. Many states, including Michigan, embraced the newfound power with gusto, establishing laws for alcohol distribution control, saloon closing times, drinking age limits and limited alcohol access on Sundays.

Hard cider consumption waned as non-alcoholic carbonated sweet drink consumption prospered. Michiganders drank water in confidence due to more reliable water systems and embraced their homegrown Faygo

Markets Bloom

D.H. Day hid this tree near Glen Haven to supply apples for moonshine during Prohibition.

and Vernors pop. As more Germans immigrated to the United States and Michigan, beer production increased.

The Great Depression of the 1930s adversely affected the livelihood of Michiganders, and many relied on gardens and orchards for food for the family. At the onset of World War II, Americans were encouraged by the government to "eat right" because "only a healthy America can win the war." Approximately twenty million Americans planted "victory gardens." In the 1940s, the United States Department of Agriculture initiated the "Basic Seven Food Wheel" as support to new dietary recommendations. The whole apple—and its fruit companions—gained momentum as healthy food options.

During this era, the Thome farm in Comstock Park on the Ridge grew its apple business. "My dad had wholesale market accounts in Detroit and Chicago," recalled Harold Thome. "He often hauled fifty bushels a load in his Model T to the Alpine [train] station. Through the 1930s, he would leave in the evening for the next day's wholesale market in Detroit."

Brian Phillips of Phillips Orchards in St. Johns recalled early family stories of fruit sales. "My great grandfather and grandfather peddled apples door to door and sold bushels to the Flint Farmers Market," said Phillips.

Family farms that survived the turmoil of the '30s regained vitality as the commercial market reignited after the war. In the Rochester area, Clayton Crissman and Henry Ross, a fruit grafter, had a surplus of apples. The fruit growers boasted of thirty-nine varieties on nearly five thousand trees between them. Varieties included Stark Delicious, Winesap, Northern Spy, Snow (Fameuse), Baldwin, Canada Red, Wagener, Shiawassee Beauty, Wolf River, Bellflower, Rhode Island Greening, Red Astrachan and more.

A large volume of apples are being shipped out of the state, but the great remaining surplus has been keeping our cider mills busy every day in the week—and the cider is mighty good too!
—Rochester Era, *October 1931*

With the fruit industry flourishing, it was especially important to establish an organization to help market the fruit. In 1939, the Michigan State Apple Commission was established to assist growers. It continues to aid them today with research, marketing and promotions as the Michigan Apple Committee.

It was very common for farms to cart fruit to the bustling Eastern Market in Detroit. Eighty-seven-year-old John Steffens recalled carting peaches and apples to the market with his dad in 1940, when he was thirteen. They traveled 170 miles one way on a bumpy route in a Ford pickup truck loaded with crates of fruit. The exhausting journey would earn them twenty-five cents a bushel for tree-picked apples and ten cents a bushel for drops that had been sorted.

More innovators launched companies to preserve fruit for the wholesale and industrial markets in the form of juicing, pureeing, drying, canning and freezing. In 1943, research began to capture the apple's fruit essence for concentrated juices and to prevent apple cider from fermenting. Fruit supply and equipment companies opened in west Michigan, including Peach Ridge Orchard Supply of Sparta in 1935 and the Fruit Picking Equipment Company of Lawrence in 1953, the latter of which developed the handy galvanized metal and cloth fruit-picking bucket.

A devastating crop year occurred in 1945 when warm March temperatures adversely affected bud growth. Just 1.25 million apple bushels were harvested, resulting in the closure of many family farms. The remaining families knew they needed to diversify to increase revenue during bountiful crop years to provide a more secure foothold in preparation of future crop losses.

Don Rasch, a fifth-generation farmer on the Ridge in Grand Rapids, shared the story that Grandpa Rob decided it was a great time to build an apple storage building and encouraged nearby farmer Walter Umlor to do the same. "You have the time; you don't have a crop," relayed Rasch. Both farmers built storage buildings with the aid of German prisoners of war (POW), who dug the foundations and received a hot noontime meal. The POWs had been sent to the Sparta POW camp in 1944 to harvest the area's crops. Since many of the Ridge families were of German descent, communication was easy between the farmers and POWs. Storage became even more important as the market changed in the postwar era. Today, Rasch Family Orchards relies on eighteen rooms to house its annual average apple harvest of 400,000 bushels.

In the 1950s, the onslaught of grocery store chains and supermarkets dampened the farmers' wholesale market revenue. Most farmers who sold fruit directly to grocery stores couldn't keep up with the demand. Supermarkets required truckloads of uniform apples that stored without bruising or turning to mush. Many of the old-time apples, grown from scions of the old apple orchards of the 1700s and 1800s, couldn't travel well for long distances and were unable to retain their beauty and shape. Some were naturally russetted with rough, brown skin and unattractive to the eye. Farmers were discouraged by the American Horticultural Society to grow apples with funny names and irregular shapes and colors. Soon dozens of apples once admired for their distinct tastes and Michigan and American heritage lost their luster. Their historical significances could not compete with beautifully colored, adaptable and transportable apples. Goodbye, Roxbury Russet; hello, Red Delicious.

Another significant development was the introduction of dwarfing rootstock. Michigan farmers learned that rootstock affected the growth and size of apple trees. Rootstock became a focus of growers, and the Dwarf Fruit Tree Association was established in 1947 in conjunction with the MSU horticultural department and the Heuser farm in Hartford. Established in 1909, the Heusers' Hilltop Orchards and Nurseries grew the state's first commercial dwarf apple tree orchard. The "dwarf tree" association, now called the International Fruit Tree Association, is an essential educational and collaborative resource.

As farmers adapted their growing techniques to better serve the commercial market, Michiganders took to the road. The convenience of automobiles provided the means for comfortable leisure travel. After World War II, accommodating destinations popped up in the form of gas stations, roadside parks, farmers' markets and root beer stands. Summer travelers responded to the welcoming diners and stopped for "burgers, fries, shakes and apple pies" on the way to breezy lakeshores.

APPLE SMORGASBORD

Since the earliest days of settlement, women on the farm have whipped up dishes starring the apple. In 1950, at the prompting of the farm wives of the Ridge, the Peach Ridge Fruit Growers Association held an outdoor event on the farm of William Schaefer featuring a delicious buffet of German-Swedish-influenced apple dishes for area families. The purpose was to showcase the apple as multidimensional—more than just for fresh eating. The success of this first farm-to-table event motivated farmers to host an invitation-only Apple Smorgasbord for the next twenty years on the Ridge. This highly attended

Opposite, top: The 1966 Peach Ridge Apple Smorgasbord at the Rasch family orchard in Grand Rapids. *Courtesy of Alpine Township Historical Commission.*

Opposite, bottom: John B. Martin, U.S. representative Gerald Ford, Michigan senator Perry Greene and hosts Royal and Bernice Klein admire the featured apple dish at the 1956 smorgasbord. *Courtesy of Alpine Township Historical Commission.*

Markets Bloom

Michigan Apples

Janet Kline, Gladys Davenport and Alma Gillett load up Apple Queen Carol Fahling's serving box at the 1953 smorgasbord. *Courtesy of Alpine Township Historical Commission.*

event attracted influential media, food market leaders and governmental officials, including U.S. representative Gerald Ford. In 1966, as first lady of the White House, Lady Bird Johnson contributed her favorite apple tart recipe. Every year, an apple queen hosted the event, which evolved to include tours, entertainment and rows of tables laden with more than two hundred homemade apple dishes. Demonstrations of apple fritter frying and copper kettle–made apple butter complemented this exclusive event.

Industry Advancements

When peach trees died after an average ten-year lifespan, more grafted apple trees were planted in their place. Demand for Michigan apples was strong, and farmers were willing to invest in their farming practices. Two significant developments in the mid- to late 1950s dramatically improved opportunities for farmers.

Markets Bloom

Until the 1950s, farmers and their workers had filled bushel crates and baskets and hauled them to the storage house and markets. With the introduction of large apple bins that held eighteen to twenty bushels, and the use of forklifts to scoop up bins and load them onto trucks, productivity dramatically increased and saved many backs in the process.

Another innovation was Controlled Atmosphere Storage, or CA, which revolutionized the industry. Before this point, apples in storage were chilled by nature and temperature control to prolong freshness. CA was developed after discovering that fruit maintained freshness in airtight storage. Apple-filled bins are stacked in large CA rooms. Once the door is sealed, oxygen is slowly reduced to roughly 1.5 percent to arrest the ripening process, essentially putting apples to sleep by slowing respiration. Temperature, humidity and carbon dioxide levels are established for each variety to maintain the fruit's freshness. When apples are needed, the door is carefully opened to fulfill orders. The apples are then delivered fresh to customers.

Between 1950s and 1970s, farmers planted orchards with dwarf and semi-dwarf tree stock rather than the standard-sized trees of old. The benefits were clear: apples gained vibrant hues and intense flavor with more uniform exposure to sunlight. Hand harvesting apples from smaller trees became much less labor intensive.

As apple orchards expanded, so, too, did the need for harvest workers. The first workers consisted of Europeans who had settled here. Families then came from the Ozark Mountain region of Arkansas and Missouri to work the orchards, followed by families from Alabama, Georgia and Florida. Later, in the 1960s, families of Mexican descent arrived, many from Texas. Julia Baehre Rothwell has many fond memories of getting to know the families and celebrating birthdays and special events: "We had numerous Mexican celebrations, including wedding anniversaries and quinceañeras." As crop yields grew, the need for skilled, hardworking harvesters increased.

Varieties grown in the orchard evolved and included an apple with Michigan roots. In Sparta in 1960, Luke Arends discovered a seedling near his McIntosh orchard bearing an early ripening apple that was pleasantly tart. He named the fruit Paula Red for his wife, Pauline. This Michigander apple grows throughout Michigan and continues to be one of the first to be picked in the harvest season. It is enjoyed fresh, as a sauce and in pie.

Also, in 1964, Phil Brown, who grew up farming on the Ridge, opened a new business to design, weld and manufacture orchard equipment to aid with farm work. Two of his innovations are the Brownie quad platform and Brownie pruning tower, used for spring pruning and fall harvests.

Orchard Markets

In the 1970s, farming operations were reevaluated as industrialization and national grocery store chains expanded into cities. Many families were tempted to sell farmland to support livelihoods. Others diversified with new on-site markets to attract customers to the farm. The Blake and Wiard families of Armada and Ypsilanti, respectively, were among the first to invite customers to pick apples in their orchards.

"We had a bunch of Wagener apples that went to Mrs. Smith pies," said Phil Wiard. "We also had thousands of Northern Spy. We promoted 'bring your own baskets' and sold bushels for one dollar, and it was a hit."

Jim Robinette of Grand Rapids visited cider mills in the southeast region and decided to "throw the dice and risk everything to buy a cider press," shared his son Ed Robinette. John Beck of St. Johns did the same. Fortunately, the gamble for the Robinettes, the Becks and many others paid off. The decision to transition from traditional farming and wholesale markets to direct selling to consumers saved many farms from closure. The era of U-pick orchards, fresh-baked pastries and pies, cider and doughnuts, wagon rides and corn mazes was launched.

Farm Wives Turn Activists

Agriculture is a fragmented industry. It needs one voice to speak and bring about effective change...It's up to farm women to offer input into problems, and we have a unique perspective. We are consumers who buy in the supermarket and we are also producers, so we can see both sides of the fence.
—*Sharon Steffens in* Michigan Farmer, *August 20, 1977*

In 1972, farm women of the Ridge and Keeler area stopped fruit truck deliveries and picketed fruit processors for underpaying farmers for apples processed for juice. Juice apples set the base price for apples used for processing into other products and for the fresh market.

The previous year, Connee Canfield, the wife of a pickle farmer, had founded Women for the Survival of Agriculture in Michigan (WSAM). Four Ridge farm women—Sharon Steffens, JoAnn Thome, Pat Cohill and Joan Hill—who quickly earned the title of "Fearsome Foursome," joined forces with Canfield to battle unjust farming laws and pricing.

Markets Bloom

We were asking $1.50 for 100 pounds of juice apples, and they were offering only $1, a price below cost of production for the fourth year in a row. We closed down the plant and got the $1.50.
—*Sharon Steffens in the* Grand Rapids Press, *January 30, 1977*

Their 1972 efforts subsequently passed the Marketing and Bargaining Act of 1973 (Public Act 344), which required growers and processors to negotiate pricing. Before long, the WSAM connected with other agricultural women in the nation to launch American Agri-Women (AAW). Steffens was appointed the first national coordinator and was responsible for editing the bimonthly newsletter. Thome became the first woman to chair the U.S. Apple Association. Together, the women coordinated and presented at numerous national conventions and met with the United States Secretaries of Agriculture to tackle tough farm issues and pass legislation that would aid farmers in continuing to grow and sell fruit for generations to follow.

Today, Steffens is active in her son's orchard market in Sparta, assisting customers with apple selections and with educational school tours.

The Dark Years

The late 1980s and early 1990s brought significant struggles when many consumers cared more for the pricing of food than where it originated. Devastatingly, between 1997 and 2000, farmers lost one dollar for every bushel of apples. The U.S. apple crops were bountiful at the same time that Chinese apple juice concentrate entered the U.S. market at prices below the cost of production. Farmers' livelihoods were jeopardized due to the culmination of events, and two hundred apple farms ceased operations.

Hooray for Honeycrisp

Then, Honeycrisp trees that had been planted in the 1990s in Michigan began to fruit. Farmers introduced this unique-tasting apple at farmers' markets, and soon it was sought by consumers, who eventually requested grocery stores to sell it. Many say Honeycrisp saved the apple industry. Tens of thousands of Honeycrisp trees have since been planted, as well as trees bearing other tasty fresh-eating varieties. The United States is home to public breeding programs connected to universities, including the University of Minnesota, Cornell University and Washington State University. Along with other countries' breeding programs, they are working hard to crossbreed apples to create new, exciting fresh eating apples. Many Michigan farmers with larger commercial orchards are planting apple trees roughly one thousand to an acre to grow apples for the fresh market. Trellising systems provide support to trees grown on dwarf rootstock that bear in five years and grow to create "fruiting walls." The compactness aids in pruning, harvesting and fruit quality.

As the Honeycrisp danced into the limelight in the late 1990s, another movement waltzed into the nation. Advocates for "slow food" encouraged people to reconnect with food grown in their own region and with America's heritage foods. This movement generated the opening of many seasonal and four-season markets in communities and at farms. Embracing fresh, local foods, people are visiting farms once again and seeking apples like Honeycrisp, Gala, Mutsu and Fuji, as well as American heritage apples like Golden Russet and Northern Spy. They are discovering many additional tasting experiences from fresh-baked goods to farm-made vinegar, sweet and hard cider, wine and brandy.

Apple Utopia

Michigan's apple crop has a significant economic impact on the state, garnering an average of $700 to 900 million in annual revenue every year. Farmers grow 900 million pounds of fruit, with 60 percent processed for other products such as sauce, juice, vinegar, slices and pies. The other 40 percent is packed and shipped to twenty-six states and eighteen countries and sold at farmers' markets. Approximately 9 million trees grow on 36,500 acres, according to the Michigan Apple Committee, with the number of trees increasing as more high-density orchards are planted. Recent bumper crops solidified Michigan's position as the third-largest grower of apples in the United States.

Between 65 to 70 percent of all Michigan apples are grown within the Fruit Ridge (formerly the Peach Ridge). About 80 percent of all fresh market apple sales are coordinated from the Ridge as well. Beyond the Ridge, 25 percent of apples grow in other counties within the Fruit Belt. The remainder of apple production can be found throughout the state, particularly in the

Lower Peninsula, in microclimate pockets ideal for apple growing. These regions benefit from ecological surroundings of rivers and bluffs, which provide the right environment for apples.

Eat a Michigan Apple a Day

Modern developments in farming practices and apple breeding have not defeated one of the greatest challenges to growers: Mother Nature. In 2012, multiple spring freeze events disastrously harmed Michigan's apple and cherry crops, resulting in a just a few million bushels. Rather than wallow, farmers and industry experts united and extensively planned and invested in new trees, new cold storage units, processing lines, innovative equipment and more wind machines to prevent future frost damage to buds. Others decided it was time to diversify again to ensure an income in the event of a future crop loss. In 2013, trees hung heavy with the weight of a thirty-million-bushel crop. In 2014, the crop was another bumper year despite a summer of unusually low sunlight.

More than two centuries strong, the commercial apple industry will continue to positively impact the state's economy. Michigan's tenacious farmers do not easily cave to tumultuous times. History has proven that these resilient, resourceful, strong-willed farmers are passionate about growing apples.

Recipes

First Lady Lady Bird Johnson's 1966 White House Apple Tarts
Courtesy of Alpine Township Historical Commission

Make a rich pastry dough (a pie crust mix may be used) and roll thin.

Cook first four ingredients together and puree when tender:

6 cooking tart apples
½ cup water
2 tablespoons butter
½ cup sugar
4 cooking apples

½ cup sugar
1 cup apricot jam
hot water

Line individual tart or small muffin tins with pastry; flute edges. Cover bottom of each tart with apple puree. Peel and slice four apples and arrange in attractive pattern on top of pureed apples. Sprinkle a little sugar over top of each tart. Bake at 400 degrees until apples are tender and crust is browned. Remove from oven. Heat apricot jam and thin with hot water to spreading consistency. Spread on tarts. Tarts may be decorated with whipped cream, cheeses or nuts. Makes eighteen tarts.

Hot German Apple-Potato Salad

Courtesy of Alpine Township Historical Commission; recipe featured at the 1974 Peach Ridge Johnny Appleseed Bicentennial

6 medium potatoes, boiled in jackets
6 slices bacon
¾ cup chopped onion
1 cup diced apple
2 tablespoons flour
1–2 tablespoons sugar
1½ teaspoons salt
½ teaspoon celery seeds
dash of pepper
¾ cup water
¼ cup cider vinegar

Peel potatoes and slice. Fry bacon slowly in skillet or electric fry pan and then drain on paper. Sauté onion in bacon fat; add diced apple. Blend in flour, sugar, salt, celery seeds and pepper. Cook at low heat until smooth and bubbly. Stir in water and vinegar. Then heat mixture to a boil, stirring constantly. Boil 1 minute. Carefully, stir in the potatoes and crumbled bacon bits. Remove from heat, cover and let stand until ready to serve. Yields 6 to 8 servings.

Chapter 3

APPLES TO APPLES

Surely, the apple is the noblest of fruits.
—Henry David Thoreau

If you've ever thrown an apple core out the window, you might have been the catalyst of a new type of apple variety. If one of the seeds from your apple toss grows into a seedling tree, it will be a new variety. This is because each seed of an apple is a unique blend of characteristics of the mama tree and the lucky pollen of papa tree, transported by an industrious bee. If a seedling sprouts, it takes up to ten years before it bears fruit, which, in all likelihood, will look more like a crab apple and taste bitter or bland. It is unusual for a seedling apple to be sublime and achieve favored status for its appealing nature. Rarer still is the seedling apple that grows and can be readily compared to its parent, though there are a few varieties whose seeds may grow into a closely matched relative. Clearly, farmers can't spread apple seeds all over their farms and hope for the best.

FROM SEED TO A (RARE) GIVING TREE

A seed hidden in the heart of an apple is an orchard invisible.
—Welsh proverb

The seedling that grows into a stately tree that bears palatable apples is a rarely found treasure. Since colonial times, America has been the birthplace of many impressive-tasting apple seedlings.

Red and Golden Delicious were auspicious seedling discoveries at the time. Red Delicious, originally coined Hawkeye, was discovered in a hedgerow in Iowa, while Golden Delicious was a lucky discovery in West Virginia. Once the Stark Bro's Nurseries and Orchards of Missouri acquired propagation rights and access to the original trees (1893 and 1914, respectively), the apples were branded with catchy, marketable names. Starting in the mid-1800s, Stark Bro's was intent on discovering new apple varieties to incorporate into their fruit catalogue and fledgling nursery business. Hawkeye was renamed Stark Delicious in 1895 and was featured on several catalogue covers. It was later renamed Red Delicious.

Today, most agree that Red Delicious has lost its original standout characteristics due to over-planting, often in less than ideal growing conditions. There are those who still seek the variety, while others praise earlier strains of the apple. Marty Brechting of Brechting's Farm Market of Comstock Park fondly recalled an older tree that he used to lay under on his family farm and munch on Red Delicious apples. The original Red Delicious strain (Stark Delicious) grows on North Manitou Island off the shore of Leland. Families can harvest one bushel a day for free, shared Kim Mann of the Sleeping Bear Dunes National Lakeshore and National Park Service.

McIntosh, discovered in Ontario, Canada, in 1811, is a seedling of Detroit's heritage apple, Fameuse (Snow), which was likely its own seedling that first grew in the St. Lawrence Valley between Montréal and Québec, Canada. Since its nineteenth-century upbringing, the dark red–forest green "Mac" has evolved into various strains, including a redder version discovered growing on a multigenerational family farm in Comstock Park. The grower, Bernard Thome, named this redder one "Ruby Mac," which has since been propagated and is widely planted throughout the United States.

To Replicate an Apple Tree

Growing an apple from the planting of a seed was common at the start of New England settlements and in New France, in which Michigan was enveloped. Soon, though, cultivated apple trees grew from scions originating in Europe and Canada. Cloning trees to propagate desired fruit is a technique used since at least 300 BC. Greek philosopher and botanist Theophrastus advocated for the budding and grafting of trees to grow the finest fruit.

Orchards bloom with acres of the same apple variety through propagation. An apple is propagated by grafting a scion ("scion wood" or cuttings) of its tree on rootstock of a young tree or older established tree. A dormant branch—the scion—with three or four buds is cut from the original tree that is to be replicated. The scion is "attached" to the new rootstock. A common grafting technique used is bench grafting in the wintertime, which entails working over a bench joining a dormant, thin scion to a one-year-old rootstock using the "whip and tongue" approach. The woods' cambium tissue, which is cut at an angle, is lined up and conjoined with wax or tape. (In the old days, thick clay, mud or horse dung was sometimes used.) The new trees rest in temperature-stable rooms of around fifty degrees. When the ground is thawed and pliable, the trees are planted. The roots then stretch to gather in nutrients, and if all goes according to science, the scion stretches and grows into a fruitful tree. The scions determine the variety, and the rootstock defines the tree size and growth vigor.

Another grafting technique is budding, which occurs in summer. A single bud collected from a leaf base is joined to rootstock to grow a new branch or tree of that bud's variety. With an old technique called cleft grafting, a scion is inserted in a split limb or stump. This technique is often used in older orchards and used to graft multiple varieties on one tree.

Since the 1800s, it has been documented that creative Michiganders have grafted a single tree with multiple varieties of apples for experimenting and for a colorful fall crop. For the most part, commercial apple growers rely on nurseries for their new trees, rather than grafting. Because commercial farmers have to order trees two to three to years in advance, some get a jump on production with grafting to replace older trees, expand orchards or supplement orchard blocks damaged by unfavorable weather. "You can put as many varieties as you want on a rootstock," said Phil Schwallier, a Michigan State University (MSU) district horticultural agent. Throughout Michigan, you can find these oddities mostly grown for fun and experimentation. The commercial grower sticks with one variety per tree to replicate popular cultivars for fresh eating and processing.

More recently, farmers are grafting trees specifically for hard cider production, like Dan Wiesen of Empire Orchards, who is consulting on a project and grafting cider apple tree varieties. Wiesen was thrilled to talk about his orchard and grafting work. "I told my wife, 'It's about time someone asked me about my apple trees,'" said Wiesen. A longtime apple grower, Wiesen is well known for his flourishing hops farm along M-72 near Sleeping Bear Dunes National Lakeshore. Wiesen's well-maintained, trellised apple orchard offers 360-degree views of National Park land. He grows Honeycrisp, Gala, Fuji, Linda Mac and more varieties for the fresh market. He is even more passionate for his grafting work to replicate varieties that provide certain characteristics to apple libations. Winter months are spent bench grafting varieties on new rootstock for spring planting on his client's farm.

To Create a New Variety

Since the advent of commercial fruit production, farmers have had to decide which fruit and which cultivar to grow to sustain farms to provide a living for their families. They made decisions on cultivars for fresh eating, baking and cooking and, as the industry evolved in the mid-twentieth century, for shelf-stable apple products like canned pie filling, apple cider vinegar and frozen, ready-made pies. Today, most farmers are planting new trees with fresh eating varieties in mind. However, 60 percent of all apples still get processed into other apple products.

At one time, roughly fourteen thousand apple cultivars were identified in the world. A tree book of the 1800s featured seventeen thousand varieties. When nurseries began selling trees in the mid-1800s, it became clear that apples had multiple names. Since then, the majority of cultivars have not survived. Apples fell out of favor because of susceptibility to disease, inability to adapt to diverse soils or survive hard winters, lackluster characteristics or traits that didn't mesh with long-distance transportation.

A historic milestone in the mid- to late 1800s was the establishment of agricultural research stations and breeding programs in the United States. The Hatch Act of 1886 prompted the availability of federal funds the following year for agricultural research. In 1889, the Michigan Agricultural College established the South Haven experimental station for agricultural research. Liberty Hyde Bailey Jr. began horticultural work at Cornell University. Throughout the United States, these collaborative research stations enabled scientists to aid growers through research testing and crossbreeding fruit for more desirable traits such as disease resistance, drought tolerance, robust flavor and skin that does not easily bruise in transport. While the South Haven experiment station focused on peach breeding, the New York State Agricultural Experiment Station (NYSAES) and other experiment stations in the United States and overseas began to crossbreed apple varieties.

Since then, tens of thousands of apple varieties have been crossbred with the hope for an offspring to be a marketable apple. Thousands of these attempts have failed in the process. Those apples that have reached the market include Cortland (1915), Empire (1945) and more than sixty others from NYSAES; Idared (1930s), developed by the Idaho Agricultural Experiment Station; and dozens of others like Fuji (1962), Gala (1976) and Jonagold (1968).

There is a possibility that Michigan will implement an apple breeding program in the future. The new Michigan Tree Fruit Commission, which gives growers greater control of research investments due to their financial input, might be the catalyst to a breeding program.

One of the most successful breeding programs in the nation is rooted in a state that doesn't rank in the top ten for apple production. The University of Minnesota established its fruit breeding program in 1878; eighty-two years later, the intentional crossing of two varieties resulted in an apple, initially called MN1711. Today, it is better known as Honeycrisp and the apple that saved a generation of farmers.

I believe that the Honeycrisp phenomenon will never be duplicated. It was a confluence of a number of factors that revolutionized the apple industry.
—Mark Doherty, industry expert

Michigan Apples Incite Breeding Career

Apple breeder David Bedford of the University of Minnesota Agricultural Experiment Station is the research scientist most connected to Honeycrisp's rise to stardom. Bedford joined the research station in 1979, five years after MN1711 had been identified for further testing. As MN1711 became established in the test orchard, Bedford continued to test and breed apples in search of a fresh, new tasting experience for the marketplace. Remarkably, it was the tasting of Michigan apples years earlier that inspired Bedford to pursue apple breeding. Raised on too many insipid Red Delicious in his school lunchbox, Bedford assumed all apples tasted unpalatable and put them in the same dislike category as Brussels sprouts and kale. After devouring half a bushel of Michigan apples brought to Bedford and other students during his undergraduate studies, he became intrigued with apples and their diverse characteristics. "Clearly, they were something better freshly harvested at perfect peak," said Bedford. And although, he does not know the variety of apples he ate, Bedford extols that Michigan bushel years ago as the impetus of his apple breeding career.

The apple breeding program at the University of Minnesota has carved a niche since Bedford's arrival for creating cold-hardy, high-quality fresh eating apples.

Developing a new apple takes twenty to thirty years of patience. "Parent" apples are selected for specific characteristics. For Bedford and his team, they consider traits that, hopefully, will blend and result in an apple that can be eaten fresh in hand, historically referred to as a "dessert" apple. Once parental varieties are selected, scientists cross-pollinate them. "We collect pollen from selected parent trees and apply it onto the flowers of other selected trees. Later in the season, we collect the hybrid seeds from these crosses and germinate them in our greenhouses. These seedlings are later transplanted into our test orchards," explained Bedford. The trees bear fruit in five to six years. If the scientists like what they taste and see, they'll take scions and graft them on rootstock to grow new trees, essentially cloning the trees.

"Every year, we discard approximately three thousand trees and find about ten promising ones that are worthy of further testing," said Bedford. "After another five to fifteen years, most of those 'promising' ones get thrown away, too." Bedford identifies flaws and favorable characteristics in each apple he creates. If a crossbred apple has interesting characteristics, it will be kept for further testing for marketability and for potential crossbreeding with another apple in the hope of capturing its good attributes.

A keen observation prevented Honeycrisp from getting tossed out. The original tree had sustained winter injury after a particularly cold winter and had been slated for removal before Bedford arrived. He decided to give the variety a second chance and waited for four young trees to bear fruit before making a final decision. These trees had been propagated from the original tree. Within a few years, the trees produced fruit, and the rest is history.

The apple tree was first planted in the 1980s in a test plot at the MSU Clarksville Research Center. After its release in the late 1990s and Michigan's early 2000s debut, Honeycrisp was introduced at farmers' markets. Soon customers were seeking it, and they were ultimately the driving force behind Honeycrisp's entry into grocery markets.

When Honeycrisp was first launched into the marketplace, its parents were misidentified. DNA testing has since confirmed that one of Honeycrisp's parents is the Keepsake. The other parent is unknown at this time. It is possible that it was an unnamed apple selection from the breeding program that had desirable traits but was not worthy of commercial release.

The Honeycrisp's phenomenon prompted breeders to focus on developing apples with unique attributes for fresh eating.

Recipes

Apple and Pulled Pork BBQ Sandwiches
Courtesy of Michigan Apple Committee

4 cups cooked shredded pork butt roast
1 cup smoky or mesquite bottled barbecue sauce
⅓ cup cider*
1 tablespoon butter
2 medium cored, sliced apples (Idared, Rome, Empire, Gala, Jonagold, Jonathan, McIntosh)
½ teaspoon cinnamon sugar
6 crusty sandwich rolls, split

Combine pork, barbecue sauce and apple cider in large saucepan. Heat over medium heat until heated through. Stir frequently. Meanwhile, melt butter in medium skillet over medium heat. Add sliced apples and cinnamon sugar. Cook and stir for 5 to 6 minutes or until apples are tender. Divide pork mixture evenly over bottom half of rolls. Spoon cooked apples over pork. Cover with tops of rolls. Yields 6 servings. (*Original recipe calls for apple juice concentrate.)

Slow Cooker Stuffing with Apples
Courtesy of Ed Dunneback and Girls Farm Market of Grand Rapids

1 cup butter, melted
2 cups chopped apples (Golden Delicious or Gala)
2 cups chopped celery
1 cup chopped onion
1 teaspoon poultry seasoning

Michigan Apples

1½ teaspoons leaf sage, crumbled
½ teaspoon pepper
1½ teaspoon salt
1 teaspoon leaf thyme, crumbled
2 eggs, beaten
4 cups chicken broth
12 cups stuffing mix

Mix butter, celery, onion, spices, eggs and broth together. Add apples and bread cubes; stir to blend. Cook in slow cooker on high for 45 minutes and then reduce heat to low for 6 hours. This recipe comes in handy when you run out of oven space at a large family gathering. The recipe yields 10 to 12 servings.

Chapter 4

Four Seasons

He who knows the apple tree knows also its region.
—Liberty Hyde Bailey Jr., The Apple-Tree, *1922*

The glaciers that swept here thousands of years ago left magnificent footprints filled with glacial waters flanking a two-peninsula landscape. Michigan's cushy location amid four of the five Great Lakes has a direct impact on the viability of agriculture. Fruit thrives particularly well along Lake Michigan's eastern shoreline from Berrien County to Antrim County. Since the mid-1800s, the region has been called the Fruit Belt. Thanks to the westerly winds of Lake Michigan and, to a smaller degree, Lake Superior, the Fruit Belt region is generally ten to forty degrees warmer than inland. The big lakes provide spring breezes that delay early blossoming to prevent bud damage from late spring frosts. Summer breezes cool evening heat, and balmy fall breezes ripen late harvest varieties. In winter, Lake Michigan protects the long shoreline from the heaviest snowfall. This "lake effect" and the region's fruit-friendly soil impact the health and growth of fruit. Farming experience and passion coax the fruit to achieve its greatest potential. To grow flavorful, high-quality apples takes four seasons of hard work.

Michigan State Apple Commission ad campaign of the 1950s. *Courtesy of Michigan Apple Committee.*

WINTER

One winter, an ice storm raged through Michigan, leaving in its wake a stunning ice mosaic. The breathtaking beauty of the arctic landscape was dampened by a closer look at tree boughs that had literally snapped under the weight of the ice. This was also the winter of record-breaking snowfall and pounding polar vortexes that wreaked havoc on pockets of orchards throughout Michigan, damaging fruit trees by the thousands. On some farms, complete apple tree blocks died under the harsh conditions. In the spring, a hailstorm bruised just-budding fruit. Collectively, though, apple orchards prevailed, with the majority of trees snug in their winter blankets. Apple trees love a good winter chill. It is their very nature to thrive in northern climates, which is why you don't eat Florida apples. Apple trees need 1,000 to 1,600 chill hours below 45 degrees Fahrenheit and can survive without any damage to temperatures as low as -25 degrees Fahrenheit.

The apple tree's winter task is to sleep. When trees are dormant, the farmer, the packer, the processor, the horticultural expert and the libation maestro (sometimes one and the same) get busy.

Every apple that was not sold at farmers' markets or shipped immediately to fresh market customers—like Meijer, Kroger or Whole Foods—is

Four Seasons

tucked away. Millions of bushels are stored in cold refrigerated rooms or CA—carefully monitored rooms controlled by atmospheric pressure.

"We have the highest concentration of on-the-farm storages in the world," said Michigan State University (MSU) district horticultural agent Phil Schwallier. "Eighty years ago, farmers marketed their own apples. They had their own small packinghouses where they sorted and packed apples into baskets and crates. Then, co-ops, like Jack Brown Produce and BelleHarvest–Belding Fruit, started taking over the marketing aspect from the growers."

As orders need to be filled, apples are taken to the packinghouses for washing, sorting, waxing, packaging and shipping nationally and overseas. Some large growers manage their apple business from planting trees to apple distribution from their own packing facilities. Small farmers may pack apples in the barn and cart them to the farmers' markets. Production ranges from roughly 2,000 bushels to 800,000 bushels a year, depending on the size of the farming operation.

John Schaefer, president of Jack Brown Produce in Sparta on the Ridge, shared that when CA storage was developed in the 1960s, "We could go to the grocery store in January and February and offer something besides tired old cabbage and potatoes. Fresh apples were just heaven. It was an absolute thrumping success." Since then, apples packaged by Jack Brown Produce labeled AppleRidge are sought year round in key markets throughout the world. Another large fresh market co-op on the Ridge is Riveridge Produce Marketing.

Approximately 60 percent of all Michigan apples are sent to processors that, in turn, peel, slice, chop, puree, pulverize and freeze apples to fulfill orders for diverse commercial products like McDonald's apple dippers, Marie Callender and Mrs. Smith pies, Indian Summer and Old Orchard apple juice, Meijer and Kroger brands, nutrition bars, taffy and pharmaceuticals. Processors include Peterson Farms, a supplier of apples to McDonald's. According to *Fruit Grower News*, McDonald's committed to buying more than twenty million pounds of Michigan apples in 2013. There are several other processors, including Cherry Growers, Cherry Central, Coloma Frozen Foods, Burnette Foods and Smeltzer Orchard.

Some growers send apples straight to processing. Most focus on growing for the fresh market. Thanks to untamed Mother Nature, though, many of these "fresh market" apples do not meet the stringent requirements of size, color and blemish-free perfection set by retailers. "Americans buy with their eyes. They like big and shiny. We can give them big and shiny," said Schaefer.

Each farm has its own strategy that, for many, evolved over multiple generations. Fourth-generation farmer Todd Fox of NJ Fox and Sons of Shelby grows an average of 225,000 bushels a year. Many of his apples go to Indian Summer for apple juice. He also grows Golden Delicious and Rome Beauty for Nestlé's Gerber brand and maintains a separate storage for them to meet the specifications for baby food. Denny Hoxsie of Hoxsie's Farm Market in Williamsburg sends fresh Northern Spy, Idared, McIntosh and Cortland apples to Peterson Farms for processing. Some are distributed to McDonald's for apple dippers, while others are sliced and packed for Brownwood Farms in the Traverse City area for apple butter. Apples are also sent to Jack Brown Produce for the fresh market and sold at the family's market on M-72 during the summer and fall seasons. For fifth-generation farmer Steve Thome, it's fresh apples: "We only grow apples for the fresh market." Even so, it is difficult to grow 80,000 bushels of perfect apples with climate curveballs and pesky pests, so many are sent to processors.

Fourth-generation farmer Dean Johnson of Johnson Orchards on Old Mission Peninsula grows a variety of apples, including McIntosh, Northern Spy, Honeycrisp, Jonagolds and Empires. The orchards encompass roughly eight hundred acres on the peninsula. Freshly harvested apples are delivered to Peterson Farms, Burnette Foods and Smeltzer Orchard for processing into apple products.

Farmers are paid for just-harvested apples when the processor collects them. On the other hand, farmers do not get paid for fresh apples until they ship, at which time the going market price for the specific apples is determined. While they get a much higher price for fresh apples than for processed apples, it might be January or later before they are paid.

Brothers Al and Joseph Dietrich manage Ridgeview Orchards in Conklin, and their cousins Jim and Mike manage nearby Leo Dietrich and Sons farm. Combined, they have a total of 1,100 acres, of which 580 are growing apple trees. Great-great-grandfather Joseph was among the first settlers to claim land on the Ridge in 1853. The orchards have greatly changed since then; however, the Dietrichs' passion for growing good fruit has not.

"We hope the interest in eating apples as a healthy snack continues," said Al Dietrich. Dietrich grows apples for the fresh market but sees opportunity in new products. A modern development is the snack-size applesauce pouch. At the start of this new market, "Cherry Growers went from four to eight lines to try to keep up with the demand," said Dietrich. His farm's apples are also processed by Peterson Farms for slicing and packaging for McDonald's, by Cherry Central for Indian Summer apple juice and by Burnette Foods for private-label juice for Meijer and Kroger. The apples are also shipped fresh to cider makers like Vander Mill for hard cider production and Dexter Cider Mill for sweet cider. To serve the growing hard cider industry, the Dietrichs have planted apple varieties that provide interesting aromas, tannins and acidity for cider. The apple industry, as a whole, constantly seeks new apple products with the aid of the Michigan Apple Committee.

The Dietrichs' farm is not a small operation. On a typically good harvest year, they hand-harvest 600,000 to 800,000 bushels of apples. That's a lot of apples. Their packing facility processes 450 apples per minute, and apples are shipped all over the world—to Israel, Central America and England, as well as within the United States. They have forty-four CA rooms that hold up to 18,000 bushels in each room.

Brothers Mike and Dave Gavin manage their 240-acre, third-generation farm, Gavin Orchards, in Coopersville from planting to packing. They've partnered with Cherry Capital Foods to distribute apples to institutions through the winter. Farm-direct to school is a key component of the Gavins' apple business, as it is for Jim Bardenhagen on Leelanau Peninsula. The Gavins and Bardenhagen bring freshly grown farm food into classrooms to share tastes and insights into where food is grown. The Gavins grow twenty-two apple varieties, including Golden Delicious, Jonagold, Granny Smith, Cortland, Ginger Gold, Gala, Honeycrisp, Mutsu and a mystery apple. Customers who visit the farm market also seek the Russet, Northern Spy and Winesap.

"Kids love the varieties we bring in," said Reneé Gavin, who is Mike's wife and a Coopersville EMT. She achieved early success providing produce to a regional college. The Gavins now supply to several school districts and host apple tasting events. They also sell "seconds" (slightly imperfect apples) at a greatly reduced price to Feeding America West Michigan and Gleaners Community Food Bank of Detroit.

Jim Bardenhagen of Bardenhagen Farms of Suttons Bay grows five and a half acres of more than two thousand apple trees. Through a partnership with the Leelanau Conservancy Farmland Preservation Program, Bardenhagen has protected his beautiful rolling farmland from future development. On his fifth-generation farm, he grows Ginger Gold, Crimson Crisp, Gala, Redfree, Honeycrisp and Chestnut Crab as a complement to his cherry orchards, potato fields and table grapes. His crops are sold fresh to regional markets, restaurants and schools. After harvests, Bardenhagen finds time to visit schools to talk about his farm crops.

Leelanau Peninsula neighbor Al Bakker of Bakker's Acres grows more than twenty apple varieties, many of which are shipped fresh into global markets, including Gala, Honeycrisp and SweeTango. His Chestnut Crab, McIntosh, Golden Russet and Spartan varieties are also popular at the orchard barn market and the Suttons Bay farmers' market. In the 1980s, Bakker spearheaded the peninsula's first farmers' markets.

The Gavins, Bardenhagen and Bakker visit schools and showcase the diversity of apple varieties and their range of tastes. It is their quest to ensure that the next generation knows apple varieties and where and how apples grow.

The Robinettes of Robinette's Apple Haus and Winery grow around fifteen varieties. They store apples in the mega "Apple Storage" refrigerator in the Apple Haus, which keeps apples cool and fresh into March. Ed Robinette said that, if needed, they can tap into apples grown by other regional farms. The Apple Haus bustles through December with fresh-baked pies and pastries for holidays and stocked shelves of bagged apples ready for a quick purchase. The Robinettes offer apple gift boxes through the holidays by mail order and press apples for unpasteurized sweet cider. Some of the cider is then fermented into hard ciders.

Besides the enormous effort to ship and process apples through winter, another massive undertaking occurs. When apple trees are assuredly dormant, pruning begins. "Every tree must be pruned every other year, though every year is preferable," said Schwallier. For large farms, "it takes three-fourths of the winter to prune."

Farmers and helpers snip or saw off tree limbs to ensure the tree's central leader remains prevalent in the subsequent harvest year. It is essential that apples get maximum sunlight for sugar content and bright hues, particularly those with a red blush or full red coloring. This effort results in a more valuable crop. Pruning will generate a higher-quality yield of apples in the forthcoming year.

Additional winter work entails seasonal worker and food safety paperwork, equipment cleaning, repairing, painting and maintenance. Winter is also the time farmers and industry experts attend educational and industry meetings, a tradition set in the late 1800s. The gatherings are essential for sharing information and learning about new challenges and opportunities.

Spring

The height of the season is planting new trees in the spring…there is so much hope for a bountiful crop and new orchards.
—*Allan Overhiser of Overhiser Orchards of South Haven*

Four Seasons

In the spring, farmers assess tree health and remove brush from rows. Conventional and organic farmers prep sprayers to fight the twenty-six major pests and diseases that can destroy orchards, including San Jose scale, Rosy aphid, obliquebanded leafroller, spotted tentiform leafminor, brown marmorated stink bug, apple scab and powdery mildew.

When the fields are dry, new orchards are planted with nursery trees. Growers who plant high-density orchard blocks install trellis systems to guide trees upward. These trees will produce within five years, with full production within ten years. The trellised "fruit walls" aid in fruit quality for the fresh market and provide easier access for harvesting fruit by hand.

In addition to planting new trees, many apple farmers grow other crops and operate farm stores. Third-generation apple and cherry farmer Suanne Shoemaker of Ed Dunneback and Girls Farm Market of Grand Rapids oversees perennial strawberry and hops fields. Daughter Stephanie opens the barn market in May to sell the spring offerings, as well as fresh-baked goods, books, kitchen gadgets and garden accessories. They also offer homemade jam and caramel sauce.

Dan and Margo Klug of Big Dan's U-Pick'em & Farm Market in Hartford plant peppers, tomatoes and cabbages by their apple orchard. Likewise, Marty Brechting of Brechting's Farm Market in Comstock Park plants peppers, kohlrabi and eggplants near his orchard.

As new plantings go in, apple blossoms unfurl. Most years, apple blossoms bloom like clockwork, typically around Mother's Day in mid- to Lower Michigan and around Memorial Day in northern Michigan. Tight clusters of buds open to reveal the center blossom, called the "king blossom." The remaining blossoms quickly follow.

As soon as the largest center blossom opens within the cluster, bees get busy. Unfortunately, for more than twenty years, there have not been enough orchard pollinators, particularly for large commercial orchards, so it is necessary for farmers to rent bee colonies from beekeepers. "Colony collapse that began in the mid-2000s and habitat loss have created an even bigger need for pollinators," explained Dr. Nikki Rothwell, MSU extension specialist and coordinator of the Northwest Michigan Horticultural Research Center. The good news is that research is underway at MSU to identify "sustainable pollination strategies for specialty crops in the United States" and develop strategies "to effectively harness the potential of native bees for crop pollination." This research is funded by a $6.9 million grant as part of the 2014 Farm Bill.

For orchards to grow fruit, hives are brought in by the truckload at night when bees are sleeping and placed in orchard rows. At sunlight, bees travel.

"Feral bees have been reduced by varroa mites and other diseases," explained Schwallier. "We rent Italian bees, though in the past few years, some have rented bumblebees." The bumblebee works in adverse conditions and at lower temperatures. The disadvantage is that it is a solitary bee that works alone. Italian honeybees are social by nature and happy to travel in packs. They prefer warm spring days and will stubbornly stay in their cozy hives in chilly spring temperatures. It is important for orchards to be flanked by nearby wildflower-filled fields and hedgerows because bees instinctively avoid cultivated fields.

As trees awaken, farmers closely monitor growth from buds to harvest. Integrated Pest Management (IPM) is used to manage beneficial and destructive pests in the orchards. Conventional and organic growers rely on IPM to determine best orchard practices from week to week. IPM scouts—MSU experts and private consultants—visit multiple orchards a day and make field observations and notes. They check traps and determine if destructive pests are present. If pest sightings are under the economic threshold, there is no need to spray. If a destructive pest or disease is above this threshold, techniques specific to those issues are instigated. Some farmers manage the IPM work themselves.

Complementing IPM reports is valuable historic and current weather data accessible via MSU's Enviro-Weather website. Temperature, humidity, wind, rain and snowfall data is entered in weather stations throughout Michigan. The weather notes and field pest observations aid scouts in determining the best tactics to maintain a healthy orchard.

If threatening spring weather is forecasted, farmers power up wind machines. The "frost fans" have saved many orchards; though, sometimes, it is just not enough.

"We ran the machines for twenty nights in the spring," recalled Don Rasch of 2012's spring hard freezes. Another hard freeze did the damage. "We had a full crop until then," said Rasch. As Mother Nature becomes more fickle, wind machines will play a bigger role in protecting spring buds and blossoms.

Dennis Mackey of Northern Natural felt he could practice organic farming well with the combination of IPM developed in the 1970s and "mating disruption" developed in the 1990s. Mackey grows organic Honeycrisps on his family's farm in Kaleva that he wholesales to Oryana in Traverse City. After a career of managing others' orchards, Mackey created an organic apple juice business. Since then, he opened Northern Natural to craft and sell wine and hard cider. Production is now managed by his son, Kyle.

Mating disruption is technically "pheromone disruption." It has been a game changer for one of the peskiest orchard pests: the codling moth. Thin bands placed strategically in the orchard at the start of the season release an abundance of synthetic pheromone to replicate the female codling moth's scent to attract a mate. The males are disoriented by "the pervasive pheromone levels and cannot find a real calling female in which to mate," explained Schwallier. This technique interrupts the reproductive cycle before damage is done. The beauty is that it does not injure or kill the pests, nor does it have an adverse effect on other beneficial pests. Destructive codling moths have been identified as a problem in the orchard since the 1800s. This innovation has saved orchards throughout the world and has also eliminated significant spraying to fight the codling moth's destructive behavior.

In May and June, the trees naturally drop fruit (and often get a little help). "Trees have the capacity to produce one hundred times more apples than we want," said Schwallier. "A dwarf tree can generate one thousand blossoms, and we only want one hundred blossoms." To grow large, flavorful fruit, "less sharing" of nutrients, sunlight and water, "the better."

Toward the end of spring, the pests become less troublesome. Even so, many farmers use creative methods for pest control, including placing hawk and bat boxes in the orchards.

"We have sparrow hawk houses at all of our farms. The hawks are great for rodent control," said Helen Dietrich of Dietrich Orchards.

"We've tried it all," laughed Don Rasch. His sister, Pat Cederholm, a historian at the Alpine Township Historical Commission, recalled her late brother Ed devouring *Mother Earth News* and trying every applicable suggestion in the orchard. "He loved dirt. He really loved the earth," recalled Rasch. In younger orchards, Ed used turkey manure compost in the spring to aid in the growth of the trees, a practice still used by King Orchards of northwest Michigan. "We order a tremendous amount of it and bring it in by the truckload," said Betsy King of King Orchards of Central Lake and Kewadin. "The trees love it. We spread it out in the fall and spring." Between the two farm locations, the Kings grow Honeycrisp, Gala, Jonagold, Ginger Gold, Lodi, Northern Spy, Paula Red and a few more varieties.

Summer

Michigan apples burst with flavor more so than apples grown in arid environments. Bright, warm summer days and cool nights combine to ripen apples with rosy cheeks and sugar content called brix.

Farmers remove suckering in late June through July and train young trees to grow upward. Some might implement trickle irrigation; others rely on rainfall and their clay and sandy loam soil.

Phil Schwallier, MSU district horticultural agent, walks the Schwallier's Country Basket orchard in Sparta to check for fruit development and pest damage.

Four Seasons

For many, July is cherry harvest, an intense two- to three-week harvest that might include U-pick days interacting with customers. A big difference between cherry and apple harvests is that machines shake cherries from trees. Apples must be carefully handpicked because they bruise easily. Beyond cherries and apples, farmers, like Allan and Kim Overhiser of Overhiser Orchards of South Haven, grow apricots, peaches and plums. If possible between harvests, farmers might escape for a quick weekend before the hectic apple harvest; though before they do, bins, ladders and storages are cleaned and prepped.

An extremely vital aspect of apple growing is skilled harvest labor. By now, farmers have communicated with seasonal workers to hire them for the fall harvest. Dawn Drake of the Michigan Agricultural Cooperative Marketing Association (MACMA), an affiliate of the Michigan Farm Bureau, shared this insight about the apple harvest work:

> *It is backbreaking, labor-intensive work that requires talent to harvest apples without bruising or dings. Most U.S. citizens do not want to do it. We have had to rely on others, mostly Hispanics, to pick fruit. If we do not have workers to help harvest the crops, we will be dependent upon imported food. Consumers will not have easy access to apples. What will be available, will be expensive. There is nothing worse for growers than having a beautiful crop and crossing fingers in the hope workers will arrive to harvest it on time.*

"Michigan is one of the few states whose growers provide licensed housing to harvest workers," said Diane Smith of the Michigan Apple Committee. Smith explained that immigration reform is needed, whether in the form of changes to the existing guest worker program or development of a new one. "Food imports mean potential food safety risks and a security risk to our country," explained Smith. "We do not want to be dependent on other countries for our food."

With a workforce secured, growers clean bins, ladders, storage units and equipment to get ready for harvest. The last of the previous year's apples has been sold. Now, it is time to start the apple harvest with the earliest ripening varieties ready in August. These vary by farm and may include tart and juicy Lodi and Paula Red. Farmers begin selling fresh apples of the season at orchard markets and community farmers' markets.

Fall

More than nine million apple trees are harvested by a gentle lift and twist of the stem or, in the case of Honeycrisp, by a gentle clip of the stem. The stem must be clipped to prevent it from puncturing neighboring apples in a bin, causing them to rot.

Once a worker has filled his or her bucket, which holds an approximate bushel, the canvas bottom is released to allow apples to gently roll into a bin. If apples are dinged, bruised, squished or other otherwise damaged, they are no longer viable fresh eating apples.

For three months, everyone works long hours every day through harvest—an exhausting effort that includes hauling apple bins to and from orchards, filling them up and carting them quickly to the packinghouse for washing, grading, waxing and shipping for immediate sales and to longer-term storage. Bins hold eighteen to twenty bushels of apples. Once filled, each bin is tagged with traceability data before it leaves the orchard row.

Bryan Bixby's three hundred acres of fruit keeps him busy from strawberry season to the fall apple harvest. Third-generation Bixby Farms of Berrien Springs is nestled within proximity to the markets of Chicago, Illinois, and Fort Wayne, Indiana, providing additional markets for apples that include Paula Red, Lodi, Ginger Gold, Honeycrisp, Gala and Cortland. Bixby manages the farm from planting crops to selling direct into fresh markets. Some of his apples are dried, mashed and powdered for fruit bar sprinkles and pulverized for use in pharmaceuticals.

Besides the enormous work coordinating with fresh market packers and processors, some farming families host school educational tours, attend apple festivals and supply apples to school districts.

Educational tours are offered by reservation. Farms around the state provide tours, including Ed Dunneback and Girls Farm Market in Grand Rapids, Erie Orchards in Erie, Erwin Orchards of South Lyon, Friske Orchards in Charlevoix, Hoxsie's Farm Market in Williamsburg, Leaman's Green Apple Barn in Freeland, Sietsema Orchards and Cider Mill in Ada and Overhiser Orchards in South Haven. As an example of a tour, Sharon Steffens and son Rob Steffens host farm tours for preschoolers to third graders at their orchard and market in Sparta. After Sharon's educational talk at Steffens Orchard Market, Rob takes the kids out to the orchard to pick an apple and see next year's bud. Oftentimes, farm tours include apples and cider.

Four Seasons

Once harvest is complete, for many farmers, it is deer hunting season. For others, it is time for a weekend nap. Then, it is time to prune the bigger apple trees and get busy again.

During harvest, the freshest tastes of Michigan apples are available in August, September and October.

Recipes

Apple Harvest Turkey Meatballs
My spin on a blend of meatball recipes using sage from my garden

1 pound ground turkey
2 tablespoons fresh, chopped sage (about 10 leaves)
1 large tart or tangy apple, grated (Empire, Northern Spy, Idared, Jonagold)
1 tablespoon finely chopped onion
1 tablespoon finely chopped sweet pepper
1 tablespoon cider vinegar
1 egg, beaten
¼ cup to ½ cup bread crumbs (for firmer meatballs, add more)
seasoning to taste (try Miracle Blend from Alden's Mill House)
¼ cup of raw cider (optional)

Blend the first 9 ingredients together and form into balls. Lightly oil cooking sheet topped with foil for easy cleanup. Place balls on sheet and bake in oven at 350 degrees for 15 to 20 minutes. After the first 10 minutes, pull sheet out and flip meatballs. As an option, pour ¼ cup raw cider over meatballs for last 5 minutes.

Apple Squash Soup
Courtesy of Big Dan's U-Pick'em & Farm Market of Hartford

1 onion, chopped
1 tablespoon olive oil
¾ cup water
1 medium butternut squash, peeled and cubed
½ teaspoon salt

Michigan Apples

½ teaspoon ground sage
14 ounces vegetable broth
2 medium tart apples, peeled and chopped (Northern Spy, Idared, Lodi, Paula Red)
1 teaspoon ginger
½ cup milk (optional)

In a large saucepan, sauté onion and sage in olive oil for about 3 minutes. Add the broth, water, apples and squash and bring to a boil. Reduce heat and cover until squash is tender, about 25 to 30 minutes. Cool until lukewarm. In a blender, process soup in batches until smooth. Reheat in the pan. Optional: add milk for a creamier soup.

Chapter 5
INTO THE ORCHARDS

> *Here's to thee, old apple-tree,*
> *Whence thou may'st bud and whence thou may'st blow,*
> *And whence thou may'st bear apples enow,*
> *Hats full! caps full!*
> *Bushel—bushel—sacks full,*
> *And my pockets full, too! Huzza!*
> *—1700s English wassailing song toasting the forthcoming apple crop with a pitcher of cyder*

It is a surprising tasting journey to compare apples to apples in any given month, especially as the season progresses and more apples become available. Early apples tend to be tarter; the longer an apple hangs on the tree, the more it is impacted by photosynthesis, which packs a sweet wallop in many October apples. There is a broader selection of apples to taste in October, though a fresh tart apple in August is a great complement to the sweet corn season.

From farm to farm and orchard block to orchard block, apple hues may vary. Location, sunlight, moisture, cool nights, seasonal weather and care all come into play on harvest day. Nature is dynamic and ever-changing, and the Michigan ecosystem is unique—like none other in the world due to four out of five Great Lakes that envelope the state. The apples of fall are fresh and perfectly tree ripened. By making treks to the orchards and farmers' markets a part of your autumnal routine, you will meet the farming families who passionately grow the apples.

Following are apple varieties grown at various orchards throughout Michigan. The descriptions are a combination of personal tasting experiences, observations and descriptions offered by the farmers. A "keeper" apple can be stored in the refrigerator or cool basement for a month or two. A few apples are seedlings of Fameuse, also known today as "Snow." As a nod to Michigan's heritage, the original nomenclature is used. Look for Jonathan, McIntosh, Northern Spy, Red Delicious and Rome in the next chapter.

Why do we need so many kinds of apples? Because there are so many folks. A person has a right to gratify his legitimate tastes. If he wants twenty or forty kinds of apples for his personal use...he should be accorded the privilege...There is merit in variety itself. It provides more points of contact with life and leads away from uniformity and monotony.
—*Liberty Hyde Bailey Jr.*, The Apple-Tree

LATE JULY TO EARLY AUGUST

One of the first apples of the season is Yellow Transparent, which I picked up at Kapnick Orchards' stand at the Chelsea Farmers Market. (Kapnick of Britton is a constant at the market and supplies apples year round.) Transparent is mouth-twistingly tart. "Add a little salt," was recommended. Instead, the plan was to roast them over the outdoor fire and then roll them in cinnamon and sugar. Before nightfall, my daughter, Makayla, who scoffs at sweet apples, ate the entire half peck of Transparents.

Robust, tart yellow-green Lodi fills the hand. An offspring of Transparent and Montgomery, Lodi was introduced in 1924 by the New York State Agricultural Experiment Station (NYSAES). When I visited King Orchards' Central Lake farm during the first week of August, a customer was excited to get this early tart variety for her first batch of applesauce of the season. King Orchards grows the apple and several other varieties at two orchards in northwest Michigan. Kim Overhiser of Overhiser Orchards of South Haven also recommends the Lodi for applesauce. It is one of the twenty-four varieties grown on the Overhisers' fifth-generation farm.

Pristine is pale yellow with a slight blush. It offers a crisp, mild sweet-tart bite. The apple is a newer variety created in 1994 by the collaborative efforts

INTO THE ORCHARDS

A quick early August Lodi harvest for a customer at King Orchards in Central Lake.

of Purdue University, Rutgers University and the University of Illinois (Purdue–Rutgers–U of Illinois) in conjunction with Indiana, New Jersey and Illinois Agricultural Experiment Stations. Scientists from these universities teamed to develop apple varieties that are more resistant to apple scab, which damages the skin. Eat Pristine fresh or use in cooking.

August to Mid-September

Spicy, sweet-tart Ginger Gold was discovered as a seedling in a Virginia orchard in 1980. It is crisp and juicy, so enjoy it fresh. Since it does not brown quickly after it is chopped, this is a great apple for salads and salsa, too. I tasted my first Ginger Gold plucked fresh from Jim Bardenhagen's orchard on Leelanau Peninsula. The apple's flavor was a surprise of pepperiness and is now a family favorite.

Seedling Golden Supreme is firm and a yellow-to-dark golden hue. Slice and dry this apple in a dehydrator or low-temperature oven. Toss it in pies and strudels. It offers a slightly sweet taste in the early season of tart varieties. It is loaded with juice, but its flavor is subtle. My husband, Kris, called it "uber juicy." Corey Lake Orchards of Three Rivers suggests this apple for cider and as a keeper as well.

The mottled blush of Mollies Delicious reflects its Gravenstein parent and makes a pretty applesauce. Introduced in 1948, its girth belies its mildly sweet taste. Chop and toss on salads and bake into pie.

Paula Red is a Michigander that was discovered by Luke Arends near his McIntosh orchard in Sparta around 1960. He named the apple for his wife, Pauline. It is a welcome early variety that is easy to enjoy fresh for its firm, juicy and slightly tart taste. It is a great backpack apple to kick off the school year, unless your child prefers apples with flavor fireworks. It makes an excellent early summer sauce. Kim Overhiser of Overhiser Orchards suggests using the Paula Red in an early season pie and for a naturally gluten-free baked apple treat:

> *Wash and core Paula Red. Leave the skin on and stuff the center of the apple with cinnamon, sugar and a tab of butter. Add granola around the bottom of the bowl, or add oats into the apple's cored center. Microwave for 1:30 minutes or until the apple is soft. Let the apple rest and then dig in for a delicious treat.*

Easy to bite, lightly juicy, aromatic Spartan was not named for Michigan State University's Sparty mascot. Rather, the sweet apple was developed in an apple breeding program in 1936 in British Columbia. It resembles its known parent, McIntosh, and bears a dark red profile.

Developed by the University of Minnesota in 1999, Zestar! earns its brand name with an exclamation point with the first bite. It is a zesty apple with distinct flavor bursts of sweet-tartness. Makayla and I visited the South

INTO THE ORCHARDS

Haven region and stopped by Husted Market in Kalamazoo. The stand was loaded with sweet corn and just-harvested Paula Red and Zestar. One bite of the apple prompted an immediate "Wow! That is zingy" from me. While Paula Red was more agreeable to my palate, Makayla loved Zestar! and called Paula Red "boring" in comparison.

SEPTEMBER

Keeper Blondee was a seedling discovered across the border in Ohio. It has a soft, lemony glow and is a newer variety that ripens around the same time as Gala and is sweet and spicy with a bit of a kick. Bushels and bags of Blondee complemented cheery red Galas at Steffens Orchard Market

Sharon Cooper stops at Steffens Orchard Market in Sparta and buys Blondee and Gala apples from Sharon Steffens for her special event.

in Sparta in mid-September. Together, they make a great midseason sampler for fresh eating.

Many a kitchen's cook has whipped up fragrant dishes with flavorful Cortland. Developed by NYSAES in the late nineteenth century and released in 1915, it bears a strong resemblance to its ancestor McIntosh. Its bright white flesh is a nod to its grandparent Fameuse as well. It is softly tart with a pinch of tanginess. It does not brown quickly, so this is another apple to chop and toss in salads.

Crimson Crisp has a smooth red-over-yellow exterior and mild sweet flavor. Like Pristine, this apple was developed by the collaborative efforts of Purdue–Rutgers–U of Illinois. Chop and toss this apple in salads and enjoy fresh.

Tuck petite Empire in lunchboxes and gym bags for a juicy, crisp, nicely balanced bite. Named for the Empire State and developed by NYSAES and released in 1945, this apple is a good keeper. Phillips Orchards' Empires get caramelized into delicious Fabiano's Caramel Apples.

Pretty, pinstriped Gala hails from New Zealand. It arrived in America in the 1990s. Just picked, it offers a snappy, flavorful bite that surely gets some of its juiciness from its grandparent Cox's Orange Pippin. Juice it up with your favorite veggies for a boost of morning nutrition. It is also excellent for drying, cooking and tossing in salads.

For many growers, Honeycrisp is among the toughest apples to grow. When the stars align perfectly, the apple excels in Michigan's cooler climate. It is a great-tasting apple to eat fresh. "Honeycrisp broke new ground for eating texture. It has a very good balance of sugar and acid. It falls in that good sweet spot, but it was the texture that broke new ground. Its bigger cells break off with a crunch and are loaded with juice," said University of Minnesota apple breeder David Bedford. One of the reasons the apple is more expensive is that it has to be clipped from the tree rather than twisted off. If the thick stem is left on, it punctures other apples in the bin. Its success in the marketplace warrants the extra care needed to grow and harvest this variety.

Macoun is a tasty little dark-purple gem developed by the NYSAES in 1932. Phyllis Kilcherman of Kilcherman's Christmas Cove Farm is a big fan of Macoun, one of her favorites of 250 varieties grown on the Northport farm. It is not a keeper, so get it when you can in late September and October. Its snowy flesh is a mirror of its McIntosh parent and is sweet and aromatic.

Redfree is a cultivar of the Purdue–Rutgers–U of Illinois breeding program. It is a smooth, aromatic red apple that offers easygoing flavors with just a touch of sweetness.

Sansa is a multicultural apple that was created from pollen of Japanese Akane and New Zealand Gala. Grower Marty Jelinek of Northport's Jelinek Orchards had plans to tap into his grafting skills, learned from his father, to replicate this beautiful, crisp, sweet apple that is comparable to its Gala parent. In the fall season, Jelinek Orchards opens a farm market at the orchard and a Suttons Bay farm stand to sell fresh apples.

Another delicious newer apple is SweeTango, which I first tasted at Charlevoix Applefest a few years ago. Two pecks were quickly purchased from Farmer's Daughter of Interwater Farms of Williamsburg. Since then, I learned that the apple was developed at the University of Minnesota and is a blend of Honeycrisp and Zestar. SweeTango is a managed variety promoted by a forty-five-member North American grower cooperative. This means that a limited number of farmers have acquired the rights to grow SweeTango. There is concern that, over time, overplanting a variety in less than ideal sites will degrade its attributes. Since SweeTango hit the marketplace, several more managed varieties have been released by breeding programs. Al Bakker of Bakker's Acres in Leelanau Peninsula grows thirteen acres of SweeTango. He is confident in his risk to invest in a newer variety in the hope that it will replicate the success of Honeycrisp. The SweeTango crop is sold globally. Locally, Bakker offers whole apples to first-time tasters at his barn market and the Suttons Bay market. Once the apple is tasted, customers often quickly return. "If you have something good, share it," said Bakker.

OCTOBER

A seedling discovery in New Zealand, Braeburn is slightly sweet and tart. Because of its firm skin, it is a great choice for salads, baking and for keeping for wintertime enjoyment. Robust and firm, it is an excellent fresh eating apple. It also makes a chunky applesauce and will hold its shape when baked in pies.

A Washington State seedling discovery of the 1980s, Cameo is a pleasing fresh eating apple. Its juicy, crisp and sweet appeal is topped off with a smidgen of tartness. Blend this flavorful apple, which might be the offspring of Red Delicious and Golden Delicious, with other varieties when baking. Keep it well into the winter in your refrigerator or cool basement.

An offspring of Red Delicious and Ralls Genet, Fuji is pleasantly low in acidity and high in sugar. Though it was introduced in Japan in 1962, the sweet apple did not arrive in the United States until years later. I had a most enjoyable rosy-flushed Fuji at Brechting's Farm Market in Comstock Park. Marty Brechting grows an earlier strain of Fuji that is firm, smooth, easy to eat, softly sweet and aromatic.

Golden Delicious is the great equalizer in our family of four. When it is more yellow than green, it is just right. It offers a nicely textured and mildly sweet-acidic bite. According to lore, when this apple was sent to Stark Bro's Nurseries and Orchards, a fledgling nursery at the time, it caught the attention of Paul Stark, who raced by rail and horseback to the tree and its owner in West Virginia. Around 1914, Stark paid $5,000 for the tree and caged and locked it to prevent others from cutting scions for propagation. The rest is history. It makes a great applesauce and is excellent in pies, other desserts and in cider. This is one of Olga Friske's favorite apples.

The very firm, very tart green Granny Smith hails from Australia in 1868. Maria Ann Smith discovered this sturdy seedling growing by her farm creek. According to the Aussie Apples website, the seedling might have grown from pomace of French crab apples that had originated in Tasmania. Not too many Michigan farmers grow this apple. You can find it at Schwallier's Country Basket in Sparta, the Fruitful Orchard in Gladwin, Big Dan's U-Pick'em & Farm Market in Hartford, Gavin Orchards in Coopersville, Phillips Orchards in St. Johns and Erie Orchards and Cider Mill in Erie, to name a few destinations.

Breeders at the Idaho Agricultural Experiment Station crossed Wagener and Jonathan, two varieties of the 1800s, to make Idared. Tangy, tart and juicy, Idared is perfect for baking because it retains its shape and is also a lip-smacking snacking apple. King Orchards has a faithful customer of the Upper Peninsula who travels to the orchard for three bushels of Idared every season. It stores well into the colder months. "Not just any old apple will see you through a yooper winter" touts the King Orchards' website. I like Idared for its crisp, flavorful bite and pretty white flesh that contrasts dramatically to its ruby red skin. Makayla liked it a lot, too, which means it appeals to those who like tart apples and those who prefer a more balanced, sweet-tangy apple.

I tasted my first Mutsu at the Charlevoix Applefest several years ago. Perfectly green, it caught my eye for its softer resemblance to Granny Smith. It is not tart like its lookalike. Rather, it offers a nice flavor balance of sugar, acid and juice with a zing. The sweetness is likely derived from one of its parents, Golden Delicious. Call it "Mut-sue" or its newer name of Crispin. This green giant is firm and an excellent keeper. Enjoy it fresh or bake it in casseroles. Angela Thompson of Nye's Apple Barn of St. Joseph combines Mutsu and Jonagold for her favorite apple pie blend.

GoldRush is tangy-tart and was developed by the breeding trio of Purdue–Rutgers–U of Illinois as a great keeping apple. Eat GoldRush fresh, bake it in pies or store for a midwinter treat.

Northern Spy and Malinda apples were crossbred to create Keepsake in 1978 at the University of Minnesota. Keepsake's offspring, Honeycrisp, gets a lot more love; however, this apple deserves its own stardom for its sweet, pleasing bite and ability to stay fresh for months in cool storage.

An apple that made its debut in the 1970s, sweet-tart Liberty is a late-season variety. It was bred at the NYSAES with Macoun as one of its parents that lends its dark red hue. It is fragrant, often compared to the McIntosh's aromas, and is divine to eat. Cook it up, press into cider or store in a cool spot to enjoy in the chill of winter.

PICK YOUR OWN

Many orchards invite you to pick your own apples during a month or two of the fall season, usually September and October. Apples ripen at different times, so when you visit a U-pick farm, know that only a few varieties will be available to pick. Contact your area farms to inquire about schedules. When you go, be sure to take great care in harvesting apples. To avoid damaging next year's bud, gently lift the apple upward and twist rather than tug the apple.

APPLE STORAGE

Always store apples in a cool place, preferably your refrigerator or in a cool basement.

Recipes

Apple Salsa
Courtesy of Michigan Apple Committee

2 medium apples (Idared, Gala, Cortland, Ginger Gold)
2 tablespoons lime juice
½ cup chopped orange segments
½ cup finely chopped onions
½ cup finely chopped green pepper
1 finely chopped jalapeno
1 clove garlic, minced
2 tablespoons chopped fresh cilantro
1 tablespoon apple cider vinegar
½ teaspoon ground cumin
1 teaspoon vegetable oil

Core and dice apples into quarter-inch pieces. Toss immediately with lime juice. Stir in remaining ingredients. Chill 2 hours. This is delicious as a dip for tortilla chips and can be served over fish, chicken or turkey. This recipe yields 3 cups.

Rose Steffens's Apple Cake
Courtesy of Sharon Steffens of Steffens Orchards of Sparta

Mix these eight ingredients together:

1⅔ cups flour
1⅓ cups sugar
¼ teaspoon baking powder
1 teaspoon baking soda
½ teaspoon cinnamon
¼ teaspoon cloves
¼ teaspoon allspice
¾ teaspoon salt

INTO THE ORCHARDS

Add the following to the mix:

⅓ *cup shortening*
⅓ *cup water*
1 *cup applesauce**
⅓ *cup nuts, chopped*
⅔ *cup raisins*

Beat for 2 minutes and add 1 egg. Beat another 2 minutes. Bake in well-greased and floured 9 x 13 pan or tube pan for 1 hour at 350 degrees. Let cool 10 minutes before removing from tube pan. Cool the cake with the right side up. (*Make your own applesauce by cooking down apples—like Cortland, McIntosh or Paula Red—with a little fresh cider or water. Puree for a smooth blend.)

Chapter 6
TASTES OF HISTORY

Through much of the twentieth century, historic, unique apple varieties with clever nomenclatures, pretty stripes and antique gold hues took a back seat to marketable, transportable apples. All for a good reason: apple farmers had to focus on growing varieties for the commercial market to maintain livelihoods.

Today, we can appreciate the best of the newer releases like Honeycrisp, Gala and Ginger Gold and still seek old-time varieties to connect us with our foodways.

The majority of Michigan's 850 apple farms grow varieties to appeal to a broad swath of the market. Still widely planted heritage varieties include McIntosh, Jonathan, Northern Spy and Red Delicious, which provide a taste of 1800s history. A handful of devoted conservators have planted apple trees that bear additional varieties with vibrant provenances. Through their work, many more Michigan heritage apples of the 1700s and 1800s are still growing in orchards today.

Just as a rose fancier would not limit himself to one variety of rose or a music lover to a single symphony, so a lover of choice fruits should be able to savor some of the infinite range of flavors, textures, colors and sizes that exists in the various varieties of apples or among grapes and even in the often neglected gooseberry.
—Robert A. Nitschke, Southmeadow Fruit Gardens 1961 catalog

Michigan Apples

Robert Nitschke planted a garden in the 1950s with unusual apple trees in the "south meadow" of his Birmingham backyard and eventually learned to graft and grow thirty varieties on one tree at one time. He eventually partnered with Theo Grootendorst of southwest Michigan, whose Dutch family had been propagating trees for more than three hundred years. The two evolved a Southmeadow Fruit Gardens catalogue and nursery to sell grafted trees of two hundred apple varieties. The business is now managed by Grootendorst's son, Peter.

Nitschke's reputation as a grower of rare trees drew admirers and customers who were inspired to grow historic apple varieties. One of those was John Kilcherman, who, along with wife, Phyllis, wished to evolve their farm's crops to include apples in 1975.

Since the Kilchermans' first planting of 3,500 trees, the varieties have been narrowed down to roughly 250 distinct apples. The Kilchermans are now sought for their expertise and devotion to antique apples. Recently, they aided the National Park Service in identifying many varieties growing within the Sleeping Bear Dunes National Lakeshore on old farmsteads of the 1800s and early 1900s. It is not uncommon for the Kilchermans to get a call inquiring about a visit to have them identify a backyard- or wild-grown apple.

In 2009, I traveled to the farm. Inside the big barn were tables laden with quarts of apples and their histories typed up on nearby note cards. The apple showcase was accented by floor-to-ceiling walls of ten thousand pop bottles. Drawn to the apples, I began learning the depths of their lively histories. Like Henry David Thoreau wrote in his 1862 "Wild Apples" essay, "I could not refuse the Blue Pearmain." After my purchase, I chomped into the russeted, burgundy skin and tasted a soft, sweet, melony earthiness—an interesting tango of flavors that, since that bite, has inspired me to seek apples of unusual color and history.

On a more recent late August visit, I admired the Kilchermans' antique apple and fruit book collection—including America's first fruit book, printed on Benjamin Franklin's Boston press—and tasted Phyllis's applesauce sampler of the early season apples. One sauce was made with Yellow Transparent, and the other was made with Duchess of Oldenburg, which had a soft pink hue to it. The tastes had slight varying notes, and I found I enjoyed the tart Transparent as an applesauce. Afterward, John and I walked to the orchard, where he grows apples that are dotted and mottled, russetted and conical—the apples are limited only by your imagination.

Like the Kilchermans, the Wards of Eastman's Antique Apples are passionate about historic apples. The Wards grow 1,500 apple varieties

Blue Pearmain hails from New England of the late 1700s/early 1800s. Find at Kilcherman's Christmas Cove Farm in Northport.

on a family farm established in 1909. A total of four thousand apple trees were first planted in 1988. Nearly two decades later, Cindy and her husband, Tim Ward, took over the farm from Cindy's dad and brother to save the trees from crop development. Together with their sons, Casey and Rafe, the Wards sell the apples at the Midland Farmers Market. The positive customer response propelled the Wards to build a charming tasting bar and production barn on the farm to craft hard cider with their colorful palette of apples that provide rich tannins, acidity and sugars that cover the taste spectrum. I combined a morning trip to the Wards' Wheeler farm with an afternoon at the Great Lakes Cider and Perry Festival in St. Johns. Driving by farm after farm of field crops, it was a delight to reach the engaging apple haven. The Wards presented a display of "antique" apples with histories spanning five centuries, including Cox's Orange Pippin, Porter's Perfection, Golden Russet and Court Pendu Plat.

Queen Charlotte was married to King George III in the mid-1700s. "The queen loved the Borsdörffer apple—also called Reinette de Misnie," said Cindy Ward, who shared its 1500s history and description as "highly esteemed for its sweet and generous flavor, and the pleasant perfume which it exhales." At one time, it was believed that the apple "dispelled epidemic fevers and madness." It first grew in the United States in 1785. The Wards have more than one thousand more apples to discover, including fifty-six cider, sixty-four russet and twenty-one red-flesh apple varieties.

In southwest Michigan in Eau Claire is a bucolic U-pick farm called Tree-Mendus Fruit that grows more than two hundred varieties. In the market, you can taste the historic apples on "showcase" days in the fall. On one such day, Bill Teichman of Tree-Mendus sliced up some samples. The varieties available for tasting in mid-September included Hidden Rose, a mystery russet and Chestnut Crab. The Hidden Rose offers a sweet-tart, juicy taste and becomingly pink interior. Belying their rustic skin, russets are really flavorful, and the mystery russet, which tasted exactly like a pear, held up to the reputation. Petite Chestnut Crab packs sweet nuttiness into a perfect little bite.

Teichman's grandfather started the family's apple history with a planting of Jonathans in 1922. Historic trees have been planted and nurtured for years by Bill's father, Herb Teichman, who originally obtained scions from the Grootendorsts' nearby nursery. The lovely orchard stretches 450 acres and encompasses a pavilion that can be rented for family reunions and other events. Plan a day to pick apples in the fall or go straight to the market for loaded-up apple crates, as many as thirty varieties at a time. Peruse the bins while you wait for your hot waffle with warm chunky applesauce topping and accompanying cider. Teichman clearly enjoys the farm life. "I get to enjoy the world's best fruit," said Teichman. "I get them perfectly ripened and the very best ones at the top of the trees."

All of the aforementioned orchards are great resources for tasting Michigan's heritage apples. There are other farms that grow historic varieties as well, like Sietsema Orchards and Cider Mill in Ada and Alber Orchard and Cider Mill in Manchester. Also find Golden Russet apples at H&W Farms in Belding and, in South Haven, at McIntosh Orchards and Overhiser Orchards. H&W also grows Sheepnose, Esopus Spitzenburg and Pound Sweet. McIntosh Orchards grows Bramley's Seedling, Pomme d'Api (Lady) and Winesap. The Fruitful Orchard in Gladwin grows Fameuse (Snow), Maiden's Blush and Gravenstein.

Tastes of History

Michigan Heritage Apples of the 1700s and 1800s

Michigan's earliest apple heritage is intricately linked to French-Canadian seeds and cultivars. The settlers ate a big dose of vitamin C with their widely planted Calville Blanc d'Hiver, which originated in France in 1539. It was likely chopped and cooked in a kettle to make tangy apple butter, sauce and flaky tarts. The apple's pronounced ribbing was made famous by Claude Monet in 1880 when he used the apple's bumpy profile as his muse in *Apples and Grapes*. Calville Rouge d'Automne originated in 1670 in France bearing the pronounced ribbing common to Calvilles. Its color can be a beautiful deep crimson, and its white flesh can be stained red from the skin.

Detroit Red is a cherry bright, compact red apple with creamy white flesh. Detroit Red is often referred to as a Canadian seedling that originated in the St. Lawrence Valley region. For a year, I worked to identify Detroit Red's origins. My best guess as to why this early noted variety was never referred to as "Détroit Rouge" is that it is really a seedling that first sprouted in Detroit—perhaps from a seed of the lost Bourassa apple. At one point, it was assumed it was the Rosseau. Historic musings also list Detroit Red and Rosseau separately when reflecting on apples of the early 1800s. Its history might be a mystery, though there is no question that it was highly regarded for its taste and bright red to dark purply skin.

Fameuse, once called "Pomme de Neige" and today more commonly known as "Snow," is the most prolific apple of Detroit's French heritage. (Shiawassee Beauty and McIntosh are seedlings.) I first discovered Fameuse at Alber Orchard of Manchester several years ago. I was taken with the apple's snowy white interior and soft spice taste. This was before I knew of its Detroit connection. Cooking up Fameuse into a batch of applesauce is a must for a snowy light sauce. It was much admired in the early days of Michigan and once described in 1874 as follows: "For an autumn apple, both on the score of beauty and excellence, it has few superiors." Unfortunately, it is an apple that is susceptible to disease, which is why it is no longer widely planted. I also found and scooped up a peck of Fameuse at The Fruitful Orchard in Gladwin and at King Orchards during Charlevoix Applefest.

Pomme d'Api (Lady) is a little bite of yellow-red sunshine that was tucked in pockets of French ladies and was popular during the French Renaissance, though its history goes way back—possibly as far as 700 BC. I often wonder if Antoine de le Mothe Cadillac brought dormant scions of the tree to Detroit in the early 1700s.

The bright red to dark purple Detroit Red was growing prevalently along the Detroit River by the early nineteenth century. Find at Eastman's Antique Apples in Wheeler.

Another tasty, albeit petite, bite is the rustic Pomme Gris (Pomme Grise), which is a russet apple. There are two theories to its provenance: it either grew as a wild seedling in Quebec or originated in 1600s France. Detroit's settlers might have tucked this keeper in their ground cellars. It offered a nutty, aromatic bite, surely an excellent complement to a meal of venison or turkey.

Land Grants and the Homestead Act Prompt Plantings

As land grant offices sold parcels to adventurers seeking land in the Michigan Territory, many brought seeds and scions from trees of New England. Following is a sampling of Michigan apple varieties of the 1800s and early 1900s.

Baldwin blasted into the spotlight in Michigan in the early 1900s. Discovered in Massachusetts in 1740, it was a superb apple for the commercial market. It traveled well and was appealing with its dotted bright red skin and tart taste. In 1921, Baldwins accounted for 25 percent of west

Michigan's apple crop. Just more than a decade later, this apple faded into obscurity. When you find a Baldwin, eat it fresh, cook it into sauce or make some cider with it.

Early to ripen, Duchess of Oldenburg originated in the eighteenth century. Its pale yellow-green skin is often covered in light red stripes. Tart, juicy and tender, it makes a tasty August applesauce, as I discovered during my visit to the Kilcherman farm in Northport.

Thomas Jefferson is believed to have favored the Esopus Spitzenburg, a flavorful fresh-eating apple that is also a favorite to use in cider. Discovered in 1790 in Esopus, New York, the apple likely lent some genes to seedling Jonathan.

Besides being fun to say, Hubbardston Nonesuch makes a great cider. It was discovered in Hubbardston, Massachusetts, in 1832. It packs a punch of fragrant flavor with a touch of acidity. Its brown-speckled, red-yellow blended exterior reminds me of the prettiest autumn treetops.

Tangy Jonathan is a heritage apple of the 1800s that is used for caramel apples and apple taffy and is a key ingredient in many sweet ciders. Some Jonathans get sent to Affy Tapple of the Chicago area for caramel apples. Jonathan is a seedling that likely grew from a seed of Esopus Spitzenburg. Its good nature was used to crossbreed other great-tasting apples. Jonathan is firm and juicy and is a great choice for sauce and pie, too. Jonagold, a cross of Jonathan and Golden Delicious, is an impressive golden-reddish, aromatic apple that is crisp, sweet, tangy and juicy. A bite of a Jonagold from The Fruitful Orchard of Gladwin had me proclaiming, "Did you see that juice fly?" Jonamac is a cross of Jonathan and McIntosh that offers a sweet-tart spicy flavor. In mid-September, look for Ruby Jon, a great baking apple that tastes tart and is perfect for applesauce.

I like all the Jonathans and the relatives, including Jonamac and Jonagold.
—Brian Phillips of Phillips Orchards

Slightly juicy New Jersey–born Maiden's Blush retains its off-white flesh color when dried, which make this blushed-yellow apple discovered in 1817 perfect for slicing up and drying in a dehydrator, a low-temperature oven or in the sun. I think my husband, Kris, fell in love with this apple at first sight, drawn to its pretty sunset glow. Its fragrance boosted its star appeal.

The Blush…though not a first-class eating apple [has] eminent beauty and unrivaled freedom from imperfections.
—Annual Report of the Michigan State Pomological Society, 1873

Hardy McIntosh, the original big Mac, was discovered in the late 1700s by Canadian John McIntosh, who determined it must be a seedling of Fameuse. Once the apple variety was propagated decades later, its scions traveled the shoreline of Lake Ontario into Michigan. It is a Michigan heritage apple of the 1800s. Betsy King of King Orchards recommends Macs for "saucy pies with lots of great flavor" and Northern Spy and Idared for "firmer pies that maintain the shape of the slices." Look for Mac in late August to mid-September.

For more than two centuries, Northern Spy (Spy) has been the go-to apple for pie, apple butter, cider and brandy. This strapping, good-looking New Yorker with a firm bearing sprouted in 1800 and was propagated more widely in 1830. "Spies for pies" is the fall trumpet song shared by growers and grandmas. Hoxsie's Farm Market of Williamsburg uses Spy for the popular caramel apple. I can imagine adventurous New Yorkers traveling by steamship and foot to the newly anointed state of Michigan with dormant Spy scions in their packs. Pomological proof of the Northern Spy's late 1800s arrival dots the Michigan landscape. Untamed Northern Spy trees are so common that when I chatted up farmers about my old childhood tree, their immediate replies were, "It's probably a Spy."

My favorite apple has always been and will probably always be the old Northern Spy. When you can find one that was picked at just the right time, or stored well until it reaches the perfect ripeness, it is like biting into a fruit that has no equal. The search for that perfect elusive Spy may be part of the fun as well as knowing that few people in the world have ever experienced that fantastic apple as it reaches its perfect time to be eaten.
—*Jim Koan, Almar Orchards and J.K.'s Scrumpy hard cider of Flushing*

The nifty thing about Porter's Perfection is that its apples often fuse together, creating little snowman-shaped apples. This 1800s apple hails from Massachusetts. It is a good cooking apple, as it retains its shape. If you like your pies firm, toss this apple into the blend.

Red Astrachan gets a lot of praise in Michigan's pomological and horticultural reports of the late 1800s for its beauty, size and overall appeal.

The Astrachan and Duchess are our two most popular and profitable market fruits for the summer. Both make fine trees…Both fruits are very handsome and fine for cooking, though scarcely first-class for the dessert.
—*Second Annual Report of the Michigan State Pomological Society, 1873*

The center "king blossom" unfurls to attract bees for pollination.

Stunning May blossoms grow into thousands of apples by fall at Lesser Farms and Orchard of Dexter.

Left: Three generations of Thome—Harold, Steve and Mitch—farm the land their ancestor settled in Alpine Township in 1846.

Below: The King Orchards crew takes a break during cherry harvest with apple bins ready for fall. Visit for U-Pick fun and apples in Central Lake and Kewadin.

A glorious sunny afternoon brightens rows of Honeycrisp trees at Blok Orchard in Ada.

Westview Orchards market and "adventure farm" was established in 1813 in Washington Township.

Ginger Gold apples, one of the author's new favorites, are almost ready for harvest at Bardenhagen Farms in Suttons Bay. *Photo by Lorri Hathaway.*

Al Bakker of Bakker's Acres in Suttons Bay in the SweeTango orchard.

Don, Nicholas and Jacob Rasch of Rasch Family Orchards manage several farms on the Ridge, just north of Grand Rapids.

Master cider maker Jim Hill of Hill Bros. Orchards and Cider Mill of Grand Rapids scoops pomace that remains after pressing apples for sweet cider.

Farmer's helpers tout tasty Honeycrisp at Charlevoix Applefest for Farmer's Daughter of Interwater Farms of Williamsburg.

Suanne Shoemaker and daughters Stephanie and Sarah of Ed Dunneback and Girls Farm in Grand Rapids offer U-Pick fun and fresh apples in the market.

Find Engelsma's cider at the Apple Barn and throughout the Grand Rapids region.

Veggie and fruit serving at Vander Mill in Spring Lake: baked root vegetable chips and Chapman's Blend hard cider.

Right: Brian Phillips manages sixth-generation Phillips Orchards and Cider Mill in St. Johns.

Below: A bin of just-harvested Jonathans at Alber Orchard in Manchester.

Opposite, top: Fresh apple pie delivery at Crane's Pie Pantry Restaurant and Bakery in Fennville.

Opposite, bottom: Northern Spy apples get a sweet topping of caramel at Hoxsie's Farm Market in Williamsburg.

Left: Dan Young and young Sadie of Tandem Ciders of Suttons Bay enjoy a sunny morning in the cider apple orchard.

Below: Chateau de Leelanau hard cider blend of Golden Delicious, Northern Spy, Honeycrisp, Idared and McIntosh grown in the Gregory orchards of Leelanau Peninsula.

Six generations have grown Overhiser Orchards to four hundred acres in South Haven. Stop by for U-Pick fun and fresh apples in the market.

Buy a bushel of Michigander Paula Red at Hanulcik Farm Market of Ionia.

Above: Cider men Eric Elliott, Bryan Ulbrich, Jay Briggs, Andy Sietsema and Mike Beck enjoy good times at the Great Lakes Cider and Perry Festival.

Left: The Ward family of Eastman's Antique Apples of Wheeler blends red-flesh apples from their orchard to craft this crisp hard cider.

Opposite, top: After a hard run, these Rockford residents enjoy hard cider at Sietsema Orchards and Cider Mill in Ada.

Opposite, bottom: More than one thousand apple varieties grow in the orchard of Eastman's Antique Apples in Wheeler.

Above: Fifth graders quickly fill a bushel worth of Fuji and McIntosh apples at Spicer Orchards in Fenton.

Left: Bill Teichman of Tree-Mendus Fruit Farm of Eau Claire slices a sample of Hidden Rose, one of more than two hundred apples that grow on the 450-acre U-Pick farm.

Just-harvested apples arrive from the orchard for washing, sorting and packing for the fresh market at Jack Brown Produce of Sparta.

Northern Spy apples grown by Wunsch Farm of Old Mission Peninsula are ready for delivery to Peterson Farms for processing into other apple products.

The sun rises on Honeycrisp at Rennie Orchards in Williamsburg.

Steffens Orchards hauls the first of many semi-loads of 1,080 bushels of fresh-picked Gala to Elite Apple packing in Sparta.

TASTES OF HISTORY

Peninsula Farmers' Club of Traverse County sent a collective exhibit of apples and pears...among the apples were Porter, Duchess of Oldenburg and Red Astrachan. The collection as a whole was very meritorious...The Red Astrachan, in the autumn of 1876, when apples were a glut in all our markets, sold readily in Chicago at three dollars to three twenty-five cents per barrel.
—*Annual Report of the Michigan State Pomological Society, 1877*

Many a Red Delicious row has been removed since the apple has lost its luster of its early heyday. Do not fret, Red Delicious fans: it is still widely grown and sold in markets for those who are passionate for this historic late-1800s apple variety. The best-tasting Red Delicious apples are mildly sweet.

Rhode Island Greening is another early American varietal that was discovered in 1650 in Rhode Island. Tart and pleasantly acidic, it is an excellent cooking and cider apple.

Rome is a mildly sweet, red charmer that grows throughout the Fruit Belt and is ripe in October. Many of its acres are devoted to supplying Gerber baby food. Rome is a luscious sauce and baking fruit. It holds its shape when baked. In the mid-1800s, Rome was discovered in Ohio by a nurseryman, who propagated the variety and sold trees to Michiganders.

Golden Russet and Roxbury Russet are brown with yellow or green undertones that reflect nature at its best in antique-gold autumn hues. Roxbury Russet is a Massachusetts native discovered in 1649. It is believed to be the first apple named in America. The apple is an excellent keeper and cider apple, but you have to eat this one fresh, too. Golden Russet tastes like honey and pear and has been around since the 1700s, though its provenance is not as well documented. The apple was a favorite of the late Fred Meijer. Another russet, named Fall Russet, was identified as a seedling that grew in Michigan. Some believe it is the Pomme Gris, the russet that first grew here in the 1700s.

In the 1800s, former peach grower George Parmalee relocated to Old Mission Peninsula in Grand Traverse County to grow apples. Parmalee remarked that the "Golden Russet is the most valuable because it is healthy, of superior quality, keeps well, does not wilt if kept in close barrels, bears well every year. It makes the best cider..." The apples sold in spring of this year when apples were plenty at four and a half to five dollars a barrel in Chicago.
—*Annual Report of the Michigan State Pomological Society, 1878*

Shiawassee Beauty is a Michigander rooted in a Shiawassee County orchard in 1850. A seedling of Fameuse, it resembles its parent in its pretty red skin and white flesh but offers a bit more spicy kick. Seeds of Fameuse were often scooped up from the cider mills of Detroit and tucked in travelers' sacks as assurance of future food for a new homestead. One of these seeds grew into a beautiful tasting, aromatic apple. The Shiawassee is more resistant to disease and bigger than its parent. It is one of the many varieties identified as flourishing on North Manitou Island in the Sleeping Bear Dunes National Lakeshore. It has been described as "vigorous and hearty" with "nature's own flavoring stored in the fruit." The apple tree might have originated in Oakland County first, before scions were grafted on trees in Shiawassee County.

No fall apple would be more esteemed—either for family use or for market purposes—than this native of our grand Peninsular State.
—Annual Report of the Michigan State Pomological Society, 1873

Tart-spicy and aromatic, Stayman Winesap (Winesap) is a 1700s cider apple that originated in Virginia, though the exact date is unknown. It is excellent in pies and a super keeper.

Tolman Sweet offers a crisp, refreshing bite. I first discovered this greeny-yellow treat at Alber Orchard in the rolling countryside of Manchester.

Wagener arrived in Michigan from New York, where it originated in 1791. This was a highly praised apple of the 1800s and might be the parent of Northern Spy. Its much-admired juiciness is appreciated in cider and sauce. Wagener holds its shape if cooked in dishes and stays fresh in cool storage for many months.

Wolf River's mammoth nature belies its mellow taste. Wolf River is a seedling discovery in 1875 in Wisconsin. Cook Wolf River into a batch of applesauce or chop and toss into any savory casserole.

The following apples were commonly grown in the 1800s orchards of Michigan for their great taste and aromas. Chenago Strawberry originated in New York. It has a strawberry-shaped profile, slightly sweet scent and pretty red blush-yellow skin. Aromatic Summer Rambo is thought to be a French apple that originated in the 1500s. It is a pretty, dotted, reddish-yellow apple. Westfield Seek-No-Further is a tangy, juicy apple that originated in Massachusetts. Winter Banana, discovered in Indiana in 1876, has a slight banana scent and is delicious fresh and in cider. Yellow Bellflower is sometimes called Sheepnose for its shape, which somewhat resembles a sheep's nose. It is juicy and flavorful. Shake this apple, and its seeds sometimes rattle in their cavities. This apple might have been the "Detroit White" apple identified as growing along the Detroit River in the early 1800s. Twenty Ounce was a heavyweight in Grand Traverse County by the 1870s. Its tart-to-sweet taste and hardiness make it a great cooking and dessert apple.

Recipe

Nutty Fruit Bars

Courtesy of Jean Rasch Piccard; featured at the 1967 Apple Smorgasbord

1 cup finely chopped apples
1 cup raisins
1 cup finely cut dates
1 cup coffee, brewed
1¼ cups firmly packed brown sugar
2 tablespoons lemon juice
1 cup sugar
2¼ cups sifted all-purpose flour
¾ cup butter
1 cup coarsely chopped walnuts
1 egg
1 teaspoon salt
1 teaspoon baking soda
¾ teaspoon cinnamon
¼ teaspoon nutmeg
1 cup buttermilk or sour milk

Combine apples, raisins, dates, coffee and ¾ cup brown sugar in saucepan. Cook over medium heat, stirring occasionally, until thick. Stir in lemon juice and cool. Combine 2 cups flour, ½ cup sugar and ½ cup brown sugar. Cut in butter until mixture is fine. Sprinkle walnuts over bottom of ungreased 13- x 9-inch pan. Sprinkle with 2 cups of the flour mixture; press down firmly. In mixing bowl, combine egg, salt, soda, cinnamon, nutmeg and buttermilk; beat well. Add remainder of crumb mixture, ¼ cup flour and ½ cup sugar; mix thoroughly. Spread fruit mixture carefully over crumb mixture in the pan. Top with batter. Bake in 350°F oven for 40 to 50 minutes, until top springs back when touched lightly in the center. Cut into

12 squares and serve with sour cream, sweetened whipped cream or ice cream as a dessert. Or cut into small bars and sprinkle with confectioners' sugar as a bar cookie.

Chapter 7

THE HUNT FOR SWEET OCTOBER

For many, a trip to the cider mill in autumn is as cyclical as the seasons, a must-do annual tradition. Crisp, chilled air and the fiery tint of the landscape prompt the drive to a favorite mill for cider, doughnuts or strudels and pies to go. For others, it is an experience discovered in adulthood. My early childhood outings included fall color tours in the Upper Peninsula. How life changes with the birth of a child. For me, it established a new tradition of annual trips to the cider mill.

In the north, we visit Friske Orchards in Charlevoix during visits to my childhood home for bags of apples, warm cider and doughnuts. We also attend the annual Applefest to stock up on apples and cider for the forthcoming winter. In the south, day trips from Chelsea include treks to Lesser Farms and Orchard for cider, apples and honey; Dexter Cider Mill for cider and apples; Alber Orchards in Manchester for apples with intriguing histories; Spicer Orchards in Fenton for U-pick fun; and Almar Orchards in Flushing for organic cider slushies and doughnuts.

SWEET CIDER

Sweet cider can be pasteurized or unpasteurized, filtered or unfiltered. Raw, unfiltered cider is juice of freshly pressed, just-picked apples from the orchard. It is thick, russeted and distinctly apple. At the start of cider season,

Michigan Apples

Olga, Richard and Jonathan Friske sell apples, cider and doughnuts at the Charlevoix Applefest, a tradition for Friske Orchards since 1985.

the juice tastes lightly tart and appears slightly more translucent. As apples are harvested throughout the season, crushed and blended together, the juice evolves into an aromatic, multilayered cider that is apple sweet and with a tart finish. The later-season cider tastes of the apple tannins—organic with complex fruit textures. Filtered raw cider is much more translucent.

Some cider makers are devoted to their specific blend. Others blend apple varieties as they are harvested. The longer the fruit hangs on the tree, the more attributes it brings to the sweet cider of October. At historic Dexter Cider Mill, you'll see a visual board with that week's specific apple varieties used to make fresh cider. Over the pressing season, they'll process up to thirty distinct apples. Other cider makers, like Phyllis and John Kilcherman of Kilcherman's Christmas Cove Farm, keep their blend hush-hush and tempt you to taste their cider made with their unique apples of antiquity.

The Hunt for Sweet October

It takes about thirty-six apples to make one gallon of apple cider. The most remarkably fresh-tasting cider is unpasteurized, preservative free and chilled. Once nature's wild yeasts eat up the sugar content, the juice ferments. Drink within a week or two to ensure you're not pouring a glass of alcoholic "hard" cider—unless, of course, that was your plan all along.

No one can dispute that raw cider beats the shelf-stable juice concentrate you find in the grocery store. Really, cider is apple juice…juice of the apple. Made seasonally, cider can be frozen for up to a year, as long as you remove an inch or two from the gallon before freezing. (This prevents a cider explosion.)

Unpasteurized, preservative-free sweet cider is a living organism that may contain harmful bacteria. Every cider producer that sells this freshest of fresh ciders labels the cider with this warning. The good news is that cider producers must follow strict guidelines to ensure a high-quality, clean end product.

"Michigan is the only state that requires an educational component for all producers to gain a license for cider making," said Mike Beck of Uncle John's Cider Mill and Fruit House Winery in St. Johns. The press at Uncle John's has evolved from a handmade unit nearly fifty years ago to one that Beck describes as a "wonderful modern example of craftsmanship and technology. We focus on fruit quality and good manufacturing practices to make cider as safe a drink as the clear juice sold in the grocery store. It essentially boils down to clean, clean, clean: the fruit, the equipment and the people [making the cider]."

If you choose to enjoy unpasteurized cider for the more true-to-the-apple taste, it is important that the apples from which the cider was crafted were picked from the trees, washed and processed in a well-maintained facility. In earlier days, "windfalls" or ground falls were scooped up from the ground and used. This has not been permitted for some time.

Regardless, pregnant women, older adults, young children and those with low immune deficiencies should not drink unpasteurized cider. And, of course, everyone has the option to seek out pasteurized cider. This cider has been heat-treated—flash pasteurized at 160 degrees for a few seconds—and then cooled down. The heat kills bacteria, but it also alters the cider's raw characteristics.

As an alternative to pasteurization, some cider makers use ultraviolet (UV) light to kill potential pathogens without raising the heat of the cider. Essentially, the cider is pumped through a tube that goes through the UV unit on the way to the chilled holding tank. "This is an FDA-approved alternative to heat pasteurization," shared Hannah Springer of Yates Cider Mill. "We think it's the best option available to ensure the safe, quality cider that our customers have come to expect from Yates."

CIDER DESTINATIONS

More than one hundred cider mills are crushing and juicing apples every fall using various styles of presses from historic wood frames to state-of-the-art steel hydraulics. For those who grew up going to the cider mill, the destinations are usually well-set traditions. My first view of a cider mill processing apples into cider was at Uncle John's Cider Mill when I fulfilled my parental duty to take my firstborn, Julia, to the mill for cider, doughnuts and farmyard fun. A stop at Uncle John's on U.S. 127 guarantees that the kids will have a great time and be zonked out in the car on the way home.

Cider mill experiences vary. Some offer elaborate corn mazes, train and wagon rides, bouncy slides, paintball, petting zoos and pumpkin patches. Destinations may charge admittance fees into activity zones.

Now that my daughters are older, I tend to seek the fresh autumnal offerings of apples and pie. Since I often find myself in Grand Rapids, Robinette's Apple Haus and Winery has become a favorite destination. The Robinettes press apples from Labor Day to early May using their fourteen-ton hydraulic press. Cider making began in the 1970s as a way to diversify the farm that Barzilla Robinette had acquired in 1911. The farm is a city respite just a few minutes from jam-packed Twenty-eighth Street and East Beltline. The Apple Haus features the bakery, cider mill and apples. The pretty farm grows a variety of fruit trees and features U-pick seasons and a winding mountain bike trail as well.

Once neighbors to the Robinettes, the Sietsemas regained their farm heritage with a new foothold in Ada. "I love to make cider," shared Skip Sietsema of Sietsema Orchards and Cider Mill. A third-generation farmer, Sietsema regrets not holding on to his family's Grand Rapids farm, though it is clear he has a passion for his new pastoral setting just a few miles east of the original homestead. A devoted grafter, Sietsema is determined to grow ugly apples that offer distinct flavors. Among the farm's 150 apple varieties are 20 kinds of russets. Sietsema plans to test them all for the best-tasting cider products. Begrudgingly, he grows hotly demanded Honeycrisp, which he sells ten to one compared to other apples in his market. His focus, though, is on apples that produce the best sweet cider and hard cider. Sietsema's son, Andy, runs the business and focuses on fermenting his cider into inventive cider libations. "We're a happy team because I love to ride the tractor and he has a good business sense," said Sietsema.

At Hy's Cider Mill in Romeo, you can watch the apple pomace pouring in the cider press from a tube and, within minutes, pouring into half-gallon jugs and being bundled-lifted into a crate for distribution to grocery stores and other markets. Grandpa's Cider Mill at Jollay Orchards offers a large viewing window of the cider press in action. At Big Dan's U-Pick'em & Farm Market in Hartford, you can stop by for cider in season and at the self-serve station. Cider is available until it runs out, usually in December. In mid-Michigan is a charming apple respite called The Fruitful Orchard and Cider Mill of Gladwin. A large window gives you a peek into the process of making delicious unpasteurized cider.

My metro Detroit friends benefit from proximity to a number of fun fall destinations. Jill recalls plenty of visits to Huron Township's Applefest in New Boston, a tradition she now shares with her two children. She recommends the doughnuts at Apple Charlie's Orchard and Cider Mill in New Boston and at Plymouth Orchards and Cider Mill, both of which are easy to reach from her home in Canton.

Tricia and Joell enjoy taking their families to Long Family Orchard, Farm and Cider Mill in Commerce Township for picking apples, wagon rides, cider and doughnuts. Cynthia's favorite is Three Cedars Farm in Lyon Township: "It's the prettiest I've ever seen, and the doughnuts are amazing." The farm brings in apples and cider through the fall season and makes a delicious apple spice cake doughnut on site. Heidi and Heather grew up visiting Franklin Cider Mill in Bloomfield Hills, which, for Heidi, still remains a family tradition, thanks to the proximity. My husband, Kris, spent many fall days as a youth sipping cider and eating doughnuts at Meckley's Flavor Fruit Farm in Somerset Center.

HISTORIC CIDER MILLS

Don't say hi, say pie.
—*Franklin Cider Mill sign*

Mills built to serve a growing population were set adjacent to waterways to utilize the natural flow of water to power machinery. As timber fell, multistory structures arose. Over time, the various floors aided in the production of cider and vinegar and leveraged the four seasons of climate, keeping cider cool on the bottom floor in the heat of summer and protecting apples from winter's frost. The 1800s structures that still exist in Michigan today are quite impressive.

Situated on the Franklin River, Franklin Cider Mill is a picturesque, rustic three-story structure that houses the cider press. The building was first erected as a gristmill in 1837, the same year of Michigan's statehood. The tranquil river setting is a few minutes from busy Telegraph Road and is testament to the various owners who used the mill for cider production since 1918. The Peltz family has owned it since 1966. Besides the delicious cinnamon spice doughnut, try a slice of Grandma Franklin's Dutch pie with a cup of cider.

Yates Cider Mill in Rochester Hills is another impressive, historic three-story structure that was originally built as a sawmill in 1863 before it became a gristmill. According to Springer, "Cider has been produced since 1876. We have press books that were kept for local farmers." On pressing days, a turbine water wheel slowly turns, providing enough power to make three hundred gallons of cider an hour. A dam diverts water from the Clinton River that travels underground via a culvert to power the turbine. The mill is adjacent to green space, perfect for a picnic of cider and doughnuts.

In mid-September, my family combined a trip to Spicer Orchards in Fenton to pick apples with a visit to the nearby Historic Parshallville Cider Mill on North Ore Creek. Parshallville opened as a flour mill in 1869 before becoming a gristmill. It was converted to a cider mill in its 100th year. Driving around the bend in the rural countryside and seeing the mill for the first time prompted an immediate, "Wow." Be sure to visit on pressing days. When you step in the retail area for your required doughnut and cup of cider, you can feel history in the bones of the beautiful building.

Dexter Cider Mill was built in 1886. When the mill was in its 100th year of operation, Richard and Katherine Koziski acquired it. In 2002, daughter Nancy Steinhauer and her husband, Marty, stepped in to own and operate the popular fall destination.

"I grew up in apple country," said Koziski, who is from the New England region. Going to the cider mill "was what you did in the fall as a youngster." This influence inspires him to seek out American heritage apples, like Baldwin and Rhode Island Greening, to add to the cider blend. "Smell is the first sense that is activated. We like apples that add a lot of aroma." Winter Banana, found at orchards that grow historic varieties, is a favorite addition to the blend. For the majority of apples, bins arrive in the early morning on pressing days from west Michigan's Dietrich Orchards. A handful of local suppliers provide supplementary apples.

In addition to aroma, Koziski seeks apples for his daughter's operation that offer the right balance of acidity, sugar, clarity and fibrous material—the

The Hunt for Sweet October

University of Michigan graduates unwind and reconnect at Dexter Cider Mill for cider and sweets.

natural tannins. The original forty-eight-inch oak rack press is still used for pressing apples. Their authentic "rack and cloth" method consists of layers of racks, each separated by synthetic cloth that holds four washed, scrubbed and ground bushels of apples. A chute from the top floor sends the crushed apples down to the press, which is below ground adjacent to the Huron River and close to inviting picnic tables. The press operation is original to the 1886 structure. Even so, the family follows the strict guidelines set by the State of Michigan for unpasteurized cider production. Watch the apples get crushed into juice, just as they were in the nineteenth century.

Near Gun Lake in Barry County is Historic Bowens Mill, which was built as a sawmill in 1838. It eventually became renowned for its buckwheat flour after it became a gristmill in 1864. The tenth owners, Marion and the late Neil Cook, bought the mammoth four-story structure in 1978 and tackled the restoration of the site after the mill had been idle for nearly forty years. In 1999, the installation of a seventeen-foot water turbine wheel was dedicated. Today, the destination is operated by the Cooks' daughter and son-in-law, Carleen and Owen Sabin. The site is home to seasonal Cider Time Festivals, which offers cider press and grindstone demonstrations, as well as a blacksmith shop and costumed Civil War camps. Visit to see history come alive, enjoy cider and pick up freshly ground cornmeal.

At a historic location in Northville is Parmenter's Northville Cider Mill, which first operated in 1873 as a cider and vinegar mill. The original structure burnt down in 1977, and a new one was quickly rebuilt to open by fall of that year. Today, its owners operate the mill from late August to just before Thanksgiving, pressing apples into fresh cider. Cider slushies, apple butter and maple syrup are just a few of the offerings.

Master Cider Makers

Jim Hill of Hill Bros. in Grand Rapids is serious about his sweet cider production. Hill is a master cider maker, one of three in the state as of 2013. Hill's cider earned three first-place awards in different years of a competitive cider competition organized by Bob Tritten of MSU Extension. His third win secured him the title of master. Ciders are judged by apple experts for "appearance and color, aroma and bouquet, acidity and sweetness, sugar-to-acid balance, body, flavor, finish and quality."

"They call me Mr. Clean," said Hill. His pressing cloths, which he hangs from a rack, are as pristine as fresh fallen snow. Hill said it is a laborious, four-step process to get the cloths pure white after each pressing. "If those aren't clean, it will show up in the product. If you want to make a good quality cider, you have to have a clean operation. Period." Hill and his wife, Arlene, make up two members of a four-person team sorting apples and sending them through the washer, grinder and vertical, accordion hydraulic press that crushes the chopped apples into juice. On a mid-September day, they processed twenty-four bins of Gala and McIntosh—roughly 432 bushels—after which Hill said they would spend another five to six hours cleaning the facility.

This attention to detail has Hill's phone ringing with orders. Hill's cider is in high demand from home brewers, including my brother-in-law Bill, and commercial brewers, who use Hill's sweet cider for hard cider production. His largest demand comes from "old German families" who travel to the orchard with their fifty-gallon bourbon barrels from Fowler and Westphalia each fall.

Hill is a sixth-generation farmer and a third-generation cider maker in the Fruit Ridge region. He knows that "Michigan apples are where it's at for hard cider." The apples provide a higher acid and more intense flavor. "The apples aren't the same away from the lakes. Washington cannot grow Jonathans, Macs or Cortlands." Hill consistently uses four apples and sometimes two more for his blend. "I have to have Jonathans and Macs. Jonathan is the cornerstone of our blend," he said. His other two reliable apples are Golden Delicious and Gala. Within the same building and adjacent to the pressing operation is an open-floor market where the Hills sell just-picked apples from their farm, fresh-pressed cider in the fall and frozen cider.

Not too far from Hill's farm on the Ridge is another cider production facility, Engelsma's Apple Barn Cider Mill, and master cider maker. Jim Engelsma gives credit to his daughter, Bridget, a part-time ER nurse, and his workers for their dedication in crafting the cider that they sell in roughly thirty Grand Rapids–area markets. When I visited, a truckload of cider was headed to neighboring farms, Wells and Moelker orchards, and another bin of half-gallon ciders was ready for delivery to Schwallier's Country Basket in Sparta. Engelsma might give credit to others now, but he drove the development of the cider business from his early days of cider making at age sixteen. "I made cider, and my friends played football," said Engelsma. His production facility is tucked behind his home in Grand Rapids and surrounded by the apple trees that supply the fruit.

"We are very fussy and run a very clean facility," said Engelsma. He and Bridget boast palates that can identify variations in the cider, including when bad or unripe apples make it into a batch. In addition to area markets, the Engelsmas' cider is sold at their nearby Apple Barn market.

The first to earn the master cider maker title were duo Bill Emery and Bill Erwin of Erwin Orchards of South Lyon, who earned the title in 2003. In addition to award-winning cider, Erwin Orchards also features wagon rides, U-pick apples, a paintball arcade and farm animals. In the market, watch freshly harvested apples turn into golden-brown liquid through a large viewing window of the press. In addition to traditional, award-winning cider is their limited-edition raspberry cider, which blends the tartness of the apple with an intense berry flavor.

Over the history of the cider competition that began in 1997, there have been many winners and honorable mentions. Plymouth Orchards and Cider Mill of Plymouth and Sietsema Orchards and Cider Mill of Ada have earned first-place honors in recent years. Klein Cider Mill and Market of Sparta earned a number of honorable mentions for Steve Klein's secret blend.

Organic Apples and Cider

Raw, unfiltered and unpasteurized cider tastes of the organic apple matter. For organic cider made from pressed organically grown apples, go to Earth First Farms in Berrien Springs, Country Mill in Charlotte and Almar Orchards in Flushing. When you visit the orchards, be sure to stock up on fresh organic apples, too. Find Northern Natural organic apples at Oryana Natural Foods Market in Traverse City.

Cider Doughnuts and Caramel Apples

Like wine and roses and peanut butter and jelly, cider and doughnuts are the ultimate pairing. My favorite cider doughnut has a crunchy exterior and soft interior that is not overly cakey. They might not be pretty, but I really enjoy Phillips Orchards' (St. Johns) craggily cider doughnuts. If you prefer fluffy doughnuts, these are not for you. Kris's favorite is the apple doughnut from Farmer White's in Williamsburg. It is fluffy and includes tiny apple bits and a fresh cider taste.

Another popular treat at cider mills and orchard markets is the caramel apple. Fabiano's Candies of Lansing blends a really smooth, easily edible caramel coating for Phillips Orchards' Empire apples. Young testers Lauren and Alex agree that the caramel apples are "awesome."

At Hoxsie's Farm Market in Williamsburg are, perhaps, the biggest caramel apples. Caramel coats robust Northern Spy and some Idared, both of which offer a nice tart accent to the sweet topping. Judy Hoxsie says they quickly sell out of caramel apples and doughnuts on the busiest fall days.

Besides doughnuts, there are many tasty apple pastries at cider mills. Bakers whip up delicious apple bread, fruit pies and muffins at Crane's Pie Pantry Restaurant and Bakery in Fennville. Olga Friske of Friske Orchards in Charlevoix peels apples for the family's baked goodies, including for her original German apple cake. The Moelker Orchards Farm Market and Bakery in Grand Rapids features apple cider doughnuts, also one of my favorites, as well as apple dumplings. Goodison Cider Mill in Rochester bakes delicious pistachio bread to pair with sweet cider.

Recipe

Apple Cider Pumpkin Spread
Courtesy of Big Dan's U-Pick'em & Farm Market of Hartford

1 can (15 oz.) pure pumpkin (or use fresh pumpkin)
1 medium apple, peeled, cored and grated (Golden Delicious, Jonagold, McIntosh, Northern Spy, Cortland)
1 cup apple cider
½ cup packed brown sugar
¾ teaspoon pumpkin pie spice

Combine pumpkin, grated apple, apple cider, brown sugar and pumpkin pie spice in a medium saucepan. Bring to a boil and then reduce heat to low. Cook, stirring now and then, for 1.5 to 2 hours. Cool. Put in a glass or plastic container that has an airtight lid. This tastes great on biscuits, toast, English muffins or pancakes. Store in the refrigerator for up to two months. Optional: add half a teaspoon of cinnamon to the mixture in the saucepan for a more cinnamony flavor.

Chapter 8

THE INTOXICATING APPLE

"It's more fun now that we're making booze with them," joked Rafe Ward of the apples grown on his family's farm, Eastman's Antique Apples. The Wards dabbled in hard cider production and garnered medals in amateur competitions for a few years before deciding to venture into the commercial market. With more than one thousand apple cultivars in their Wheeler orchard, the Wards have plenty to tap into for their hard cider production.

The Wards are in good company. Since the mid-1990s, there has been a rebirth of the use of apples (as well as cherries, pears and other fruits) for sipping pleasure beyond sweet, non-alcoholic refreshments. And while fresh-to-market apples must meet stringent commercial standards for size, color and good looks, fermented and distilled apples do not need to be the prettiest. They just have to offer the right characteristics.

History shows that Michiganders have always liked their drink. In post-Prohibition times, there were a few particular milestones in Michigan that nudged the apple along into its current buzz-worthy role as headliner of hard cider, wine, spirits and cyser.

In the early 1990s, Philip Korson II, of the Cherry Marketing Institute, and Kris Berglund, PhD, of Michigan State University, researched the viability of distilling Michigan's plentiful fruit resources into brandies, also called *eaux de vie*, in the style of Germany's *kirschwasser* (cherry brandy) and France's Calvados (apple brandy). By 1996, their leadership had driven the overturning of a Prohibition-era law and made micro-distilling licenses affordable for the production of fruit brandy below sixty thousand gallons a year. To get a license, one must have a still. Korson, Berglund and a handful of Michigan winemakers traveled to the Black Forest region of Germany, where, within a week, they had ordered Christian CARL hand-pounded copper and stainless steel stills.

In 2008, the passing of Public Act 218 propelled distillation efforts by reducing fees again and enabling micro distilleries to craft any type of spirit under one license and sell a glass of spirits in tasting rooms, crucial for educating people on how to best enjoy the spirits and to, ultimately, spur sales. To be as clear as an *eau de vie*: the apple spirits crafted today differ vastly from the volatile front-stoop applejack of pioneer times.

Black Star Farms in Suttons Bay uses thirty to forty pounds of fruit to make one bottle of *eau de vie*. The resulting spirit is ideal for sipping as an aperitif or for flambéing or blending in culinary dishes. Black Star was one of the first in the state to distill fruit under the new license. For the Spirit of Apple, winemaker Lee Lutes experiments with multiple barrels to add varying toasted undertones. To achieve interesting layers of flavor, he blends distilled fruit from both young and aged barrels. Once aged, the spirit is blended for depth with an emphasis on mellowed sweetness, vanilla undertones and ripe apple essence.

Lutes and his Black Star Farms' team are among several artisans who extrapolate flavors in more than one way. Apple dessert wine, hard cider and cherry hard cider are part of the Black Star collection, which are made with various apples of the region such as Winesap, Northern Spy, Rhode Island Greening, Jonathan, Golden Delicious, McIntosh, Gala, Empire and crab apples.

Due south a couple hundred miles is another destination where *eaux de vie* have been crafted since the late 1990s. The late Dayton Hubbard of Corey Lake Orchards of Three Rivers had been inspired by his upbringing in Floyd County, Virginia. Hubbard grew up in a one-room house without running water "on a road without a name that connected to another road without a name…that connected to a road that did have a name," shared his granddaughter Becca Sonday. Hubbard worked with his orchard neighbor Bruce Ruesink to develop more than ten styles of brandy, including the most popular apple brandy. With Hubbard's passing in 2014, the farm is now owned by Sonday's mother and aunts. Sonday is working with Ruesink to continue the tradition of distilling estate fruit and has plans to offer hard cider and cyser, the latter using honey collected from beehives managed by Sonday and her mom.

You can't talk about apple libations without talking about Mike Beck. Beck is a fifth-generation farmer at Uncle John's Cider Mill and Fruit House

Winery in St. Johns. If it involves apples, Beck has crafted it. Adjacent to his market and cider mill is a tasting room with racks of Beck's apple creations: wine, brandy, vodka, hard cider and cyser.

Beck helped launch the apple and its juice to a loftier sip. While he made apple brandy on the farm in his still, Beck teamed with Berglund, a distinguished professor of chemical engineering at MSU, to develop the state's first apple vodka using MSU's special vodka still and Beck's fresh apples. The vodka still has moved since the first production year to Red Cedar Spirits of East Lansing, which Berglund uses as a teaching and research and development facility and tasting room.

"Our vodka makes the most refreshing vodka and tonic," said Beck. "It also adds a unique layer of complexity to a Moscow Mule." For the latter, pour the vodka in a copper mug or chilled glass. Squeeze lime and drop in the liquid. Pour in a ginger beer.

Many micro distilleries, using fruits and grains, have opened as standalone businesses. This industry will continue to grow as more farms diversify.

Interestingly, the craft brew scene emerged around the same time as the micro-distilling industry. Eventually, interest in craft brews leapt ahead, thanks to smaller batch microbrews that blended hops with surprising ingredients. The industry soared and opened the door for fruit-based spirits and craft hard ciders as more people became interested in local libations with unique flavors and gluten-free options.

Hard Cider

Give me yesterday's bread, this day's flesh, and last year's cyder.
—*Benjamin Franklin,* Poor Richard's Almanack

"He's definitely one of the Jedi knights," said Jay Briggs of Forty-Five North in Lake Leelanau in reply to my prompt that, perhaps, Mike Beck is the Yoda of the hard cider industry. Beck's positive influence is widespread, and hard cider artisans who have crafted ciders for several years and those entering the market find Beck a great resource and encouraging industry leader.

"I try to help people by telling them about my experiences and letting them see how I do things," said Beck. "There are no secrets to what I do. The fruit quality, above all, will make folks successful."

MICHIGAN APPLES

In the late 1990s and early to mid-2000s, the modern-day hard cider industry was launched as a natural diversification of farming. As a board member of Michigan Apple Committee in the late 1990s, Beck encouraged MSU to pursue grant funding for hard cider research.

"The apple committee was looking for all possible markets for Michigan fruit," explained Beck. "We needed to keep up the high demand." This was the era when apple juice concentrate from China entered the domestic market and many growers struggled to maintain family farms with livelihoods strongly tied to the apple juice industry. For Beck, he wanted to generate revenue when his retail markets closed during the off-season. His hard cider is now available in cans and distributed throughout Michigan and into other key markets, including Chicago.

McIntosh Orchards in South Haven offers draught hard cider blends, fresh apples and a pretty orchard view.

THE INTOXICATING APPLE

Bruce McIntosh applied for a federal license to craft hard cider and wine in 1998, but when it was time for a facility inspection, he replied, "Well, it's not built yet." This misstep in the licensing process pushed the opening of McIntosh Orchards to 2007, when he opened a lovely farmhouse tasting room on an old fruit orchard in South Haven. It was a dream fulfilled for McIntosh, who attributes fond memories of riding his bike as a child to a cider mill near his New Jersey origins as the impetus for his tasting room in retirement. "There's too much hurrying. Slow down," encourages McIntosh. "The new 'American Gothic' is to unwind with mugs of cider." This is easy to do with McIntosh's dry and semi-dry hard ciders, five-apple blends that include McIntoshes.

John Linardos of Motor City Brewing Works in Detroit began making a traditional dry hard cider in 1998 as a nod to Detroit's apple heritage. To keep up with demand, Linardos and his partner, Dan Scarsella, a fifth-generation Detroiter, obtain apples from west Michigan orchards. Together, they've located and mapped Detroit's remaining historic apple trees grown on private properties, abandoned lots, city parks and roadsides. Each fall, they collect fruit from these old trees, often a copious amount of Baldwins and crab apples, and make a few ceremonial batches of hard cider for the taproom and special events.

Cider was the quintessential drink of our country's founders and Michigan's earliest settlers and pioneers. For some Michigan farmers, like Jim Koan, hard cider has just always been. His J.K.'s Scrumpy Orchard Gate Gold hard cider is made from apples grown on trees planted in the same organic orchards his great-great grandfather planted in the 1850s. According to Koan, "Cider saved the farm during the Great Depression" and was a highly sought "special farm cider" during Prohibition. As natural as the sun rising and setting, apples were grown, cider was made, fermentation occurred and hard cider was savored after an arduous day. Koan is a modern-day, global Johnny Appleseed, spreading the transcendent taste of the fermented apple. Koan succeeds well beyond Johnny's old territory, though. He works through a distributer to sell J.K.'s Scrumpy ciders in niche retailers in the United States, Canada and the Pacific Rim.

Commercially, hard cider disappeared from the American landscape after Prohibition. With hard cider at the forefront again, it is important to understand its historic demise. As stated earlier, John Chapman (aka Johnny Appleseed) does not get credit for Michigan's first orchards that bloomed along the Detroit River in the 1700s (and subsequent drinking of cider), though he deserves a heartfelt cheers and toast. The strolling chap played a critical role in America's pioneer movement to settle the West beyond the original colonies. Throughout the Ohio Valley, Chapman's seedling trees aided settlers of the late 1800s and early 1900s, who were required under the Homestead Act to improve the land they claimed. Chapman was one of America's first nurserymen who intentionally scooped up apple pomace at mills and planted seeds along his journey. Once seedlings grew, he often sold them or gave them away, essentially jumpstarting homesteads and hard cider and applejack production for many of America's earliest pioneers. Chapman's trees grew from seeds that produced new, unique apples, largely of a tart, acidic profile. Once crushed into juice, the apples' wild yeast naturally fermented into an alcoholic beverage enjoyed by all, even young children, who sipped ciderkin, a slightly diluted fermented cider. It is likely that some of Chapman's seedlings eventually made their way to Michigan after statehood, safely tucked in satchels of adventurers trekking northward from Pennsylvania, Ohio and Indiana.

Drinking water was questionable until public water systems evolved, so naturally settlers quenched their thirsts with a mug of fermented fruit. Along with fresh-caught fish, cider was a popular export from Detroit's harbor in 1818, generating a combined $60,000 in sales. By the mid-1900s, a new scene unfolded. Most definitely, Michigan's two prohibition periods aided in the demise of hard cider. However, there were many other factors that wiped cider from the patchwork of America. Sure, axe-swinging temperance activists chopped down some apple trees as an attempt to deter the production of hard cider and brandy. But let's get real: there were plenty of hooch-chugging happy hours still happening in barn cellars.

A string of happenstances adversely impacted the pioneers' drink. In the 1840s, Germans and the Irish arrived in America with beer and whiskey. American beer evolved into hardy German-style lagers that could be brewed anywhere. Hard cider, on the other hand, was made in the orchards. As rural living decreased in direct relation to increased urban living, settlers sought libations that could be easily and cheaply made in the city. As the population became more mobile due to improved roadways and the introduction of the automobile, there was a greater ease to transporting beer. After Prohibition, cider could not legally contain preservatives, whereas wine and beer could be made with preservatives and were thus more transportable over long distances. As America's profile became a melting pot of cultures, hard cider faded from the scene. Until now.

Barzilla Robinette discouraged the family from naming future generations after him. His great-grandson Bill paid tribute more creatively with his first

The Intoxicating Apple

Barzilla Hard Cider release. Robinette's Apple Haus and Winery is tucked in a woody, elevated bend in Grand Rapids. In the lower-level winery tasting room, you can sample wines and hard ciders, which also feature cherry and cranberry. Fermenting fruit was a natural evolution of the family farm.

It was 2006 when I first tasted Robinette's Barzilla and Beck's Farmhouse Cider. Both surprised me with their fresh, crisp apple taste. It has been fun to reconnect with the ciders and taste new ones as well. A lucky discovery was J. Trees Cellars' Farmstead Cider. For that batch, cider maker Jon Treloar used Golden Delicious, the only apples he could get during the 2012 apple shortage. It tasted crisp and appley. Others that taste of just the apple include Vander Mill's semi-dry Chapman's Blend and Left Foot Charley's Relic.

Andy Sietsema of Sietsema's Orchards and Cider Mill and Jay Briggs of Forty-Five North offer traditional hard ciders and ones that combine hops and fermented apples for microbrew fans. Briggs describes his dry hops cider as "really aromatic and grassy with citrusy characteristics." He uses Citra, which he calls "the Sauvignon Blanc of hops." Sietsema makes a dry hop cider, too, with four different hops for an earthier palate punch.

While Briggs collects apples from northern Michigan, Sietsema taps into his family's Ada farm apples, which offer plenty for small-batch production. A total of 150 apple varieties are planted in the soils. Unique characteristics are offered by such old-time varieties as Ashmead's Kernel, red-flesh Niedzwetzkyana Crab, Golden Russet, Harrison, Roxbury Russet, Esopus Spitzenburg and Newton Pippin.

"I want my hard ciders to be as natural as can be, although sometimes it's necessary to add yeasts and tannins," said Sietsema. Sietsema also offers a hard cider that is aged in bourbon barrels and another that has lemongrass added to it. He frequently hosts summer and fall cider dinners and, since 2014, a "hard cider run" through the orchards.

Just like with wines and brews, it is fun to taste the different hard cider styles. Briggs crafts a wild-fermented cider, aptly called Wild Cider. He lets the wild yeast "get funky" and takes the apple blend's juices wherever it wants to go. "Each batch will be different and reminiscent of gritty Normandy-style ciders," said Briggs.

Many hard cider makers, including Sietsema and Briggs, praise the leadership of Dan Young and Nikki Rothwell of Tandem Ciders. There is a lot of synergy in the business, and that can be attributed to those like Beck, Young and Rothwell, who are more than happy to share knowledge and experiences. In 2014, they traveled with other U.S. cider makers to Costa

Verde of Spain to get a peek into one of the oldest (600 BC)—Spaniards claim the oldest—cider production regions of the world.

Dr. Rothwell is a MSU extension specialist and the coordinator of the Northwest Michigan Horticultural Research Center. Her expertise and Young's background in brewing result in a symphony of ciders that Young describes as "expressions of the apples we grow here." This isn't to say that they don't offer a traditional English style. Crabster is just that. A dry blend of cultivated apples and foraged, tannic-rich cider apples, it is more of a traditional farmhouse cider. It was this style of cider that Young and Rothwell discovered while traveling by tandem bicycle along England's southern coast with the intention to sip many mugs of ale. Since opening the Tandem Ciders tasting room on their charming Leelanau Peninsula farm, they have developed more ciders that reflect Michigan's fruit heritage, like lip-smacking Smackintosh, made with an especially sweet-tart blend of McIntosh, Rhode Island Greening and Northern Spy from northwest Michigan farms.

Regarded as oddities in the not-too-distant past, many apples of antiquity—those with profound provenances and, oftentimes, unusual names—are providing cider makers with flavor notes, texture and acidity that cannot be found in more widely planted varieties. Admired for years for their 250 apple varieties, John and Phyllis Kilcherman of Kilcherman's Christmas Cove Farm in Northport are suddenly in even greater demand for their apples as part of the hard cider scene.

Likewise, 1,500 apple varieties at Cindy and Tim Ward's Eastman's Antique Apples farm in Wheeler are sought for hard cider production. The Wards have supplied apples to the cider makers of Tandem Ciders, Uncle John's, J. Trees and Motor City Brewing Works. Now that the Wards are selling commercial hard cider, it is getting tougher to distribute apples. Demand is far exceeding supply. Among the Wards' early releases are red-hued hard ciders, crafted from red skin-to-flesh orchard varieties. "Hard cider apples give the cider a very unique taste that comes from the tannins that these apples possess," said Cindy Ward. Visit the Ward's Wheeler tasting room for hard cider, fresh apples and pretty farmland not too far from Midland and Alma.

Steve Lecklider manages his family's farm, Lehman's Orchard, which was established in Niles in 1929. A surplus of apples in 2008 prompted him to craft hard cider. Since then, his libations have evolved to include apple brandy and wines using estate-grown fruit. Lecklider has honed his grafting skills to propagate trees bearing heritage apple varieties like Fameuse and Esopus

Spitzenburg, which he sells at his farm. He also has planted three acres of heritage apples, including Rhode Island Greening and Roxbury Russet, as a complement to his acreage devoted to fresh market apples such as Honeycrisp. Lehman's Tap House, a tasting room featuring Lecklider's libation creations, including wines and brews, is his newest venture in Buchanan.

You could say that cider makers are helping Michiganders reconnect with long-forgotten varietals, many of which grew in America in the 1700s to early 1900s. Cider makers are snipping off scions of cultivated and long-forgotten apple trees bearing apples that provide acid and tannins perfect for balancing cider flavors. They are using old-school grafting techniques to propagate varieties. The young trees are being planted on farm plots in the hope of vigorous crops.

As a complement to fresh eating and sweet cider apples, Beck planted English hard cider varieties such as Kingston Black, Dabinette and Yarlington Mill; French varieties such as Bedan and Domain; and American varieties such as Wickson, Esopus Spitzenburg and Winesap.

Ben Crow of Crow's Hard Cider and Good Neighbor Organic and Kyle Mackey of Northern Natural share a few things in common. They both offer organic hard ciders and plenty of flavor twists. Crow is proud of his award-winning Ciderye, a thirty-proof organic cider aged in rye whiskey barrels for two years. He has a deep passion for French-style ciders that are semi-sweet with a little effervescence, as well as for cider infusions. His earlymorning pickup is a coffee-infused cider he created with locally supplied organic beans and estate and Leelanau Peninsula–grown apples.

Mackey is foremost a chef. His culinary background inspires infusions of lavender, blueberry, elderberry, cranberry and other bright flavors to expand the apple sips offered in the tasting rooms of Traverse City and Kaleva. His father, Dennis Mackey, started Northern Natural in 2007 as his opus of a four-decade career managing orchards and growing organic apples.

Crow's Northport tasting room boasts a moose head, fieldstone fireplace and casual vibe. "I'm all about tattoos, T-shirts and sandals. I pour booze in your glass, you drink it and then you buy it."

For apple farms, diversity is becoming even more important to continued longevity. The 2012 crop failure jolted many into implementing strategies for diversification. The Blake family of Blake Farms, established in 1946, added a tasting room on the family's popular farm destination in Armada to offer ciders and wines year round. Steve Meckley of Meckley's Flavor Fruit Farm, established in 1956, added hard cider production to diversify as well. Hard cider is one other product that helps sustain the family farm.

Rob Hagger of Crane's Pie Pantry Restaurant and Bakery evolved his family's Fennville farm business in 2014, just two years shy of the farm's one-hundred-year anniversary. Leaving his teaching career behind, Hagger took over management of the destination and hired experienced libation maestro Eric Heavilin to craft high-quality, small-batch hard ciders, wines, meads and cysers. Heavilin, formerly with Fenn Valley Vineyards, offers years of experience producing hard cider and wine for breweries and wineries around the state. At the tasting bar inside the restaurant is a novel culinary concept. Order a custom flight of Michigan fruit desserts to accompany your flight of hard ciders and wine. Taste hard cider with an apple crisp or blueberry hard cider with a blueberry strudel. Tempting, isn't it?

APPLE WINE

The apple wine is the star attraction at the tasting room at the Fox Barn Market and Winery in Shelby. Winemaker Kellie Fox ferments apple juice from local cider maker Jason Fleming, who presses apples from regional farms, including a small amount from NJ Fox and Sons, run by Kellie's husband, Todd, a fourth-generation farmer. Fox makes Harvest Apple, which she describes as "a sweet apple wine with a kick." Fox recommends adding mulling spices to a warm crockpot of Harvest Apple for an autumn treat. A self-proclaimed frugal winemaker, she makes just enough wine to carry over until the next fall crop, which means that each bottle is a fresh taste of the previous harvest.

In most cases, if hard cider is on the menu, you'll find an apple wine, which has been fermented and coaxed into a higher alcoholic drink that is enjoyed for its smooth fruit-forwardness. Whereas hard cider is generally in the 5 to 7 percent ABV (alcohol by volume) range, wine is typically 10 to 14 percent ABV. To compare, Black Star Farms' Spirit of the Apple is 40 percent ABV.

Bruce McIntosh of McIntosh Orchards makes a Golden Russet apple wine from trees he planted in 1999. "It parallels a Riesling," said McIntosh proudly. "It's sweet and crisp and has even won a few medals." The apple's history began in 1845 as a New York seedling, and it is a favorite to ferment.

Apple wine is also available at Northern Natural, Robinette's, Royal Farms and Spicer Orchards.

CYSER

In the 1800s and early 1900s, it was common for apple farms to manage a bee colony of hives to supply ample pollination for fruit trees and to collect honey, an essential farm ingredient. Some apple farms manage hives today, making cyser a natural product evolution. More often, though, honey is collected from nearby beekeepers. The fermentation of honey is an ancient discovery, so it was only natural to ferment honey and apples together. Cyser is a surprising taste, a bit thicker on the palate than a hard cider. It can be crafted as a sweet to semi-dry libation. B. Nektar, a mead house in Ferndale, crafts Zombie Killer, a carbonated brew of fermented apples, honey and tart cherries. Mike Beck crafts a cyser as well, a tribute to his grandfather who operated a bee business until the late 1950s and supplied beeswax to the U.S. Army during World War II for waterproofing and polishing and to supplement the sugar ration. A renewed interest in mead and cyser means more options on the libation tasting menu.

RECIPES

Applesauce and Pears
Grandma Schneider's recipe; courtesy of Amy Jo Dayton of Traverse City

5 to 6 pounds of apples and pears
½ bottle of bourbon
¾ cup of brown sugar
raisins (optional)
pinch of salt
¾ cup of white sugar
½ a lemon
½ cup a water

Chop up apples and pears and soak in a half a bottle of bourbon with ¾ cup of brown sugar. (Raisins can be added to the fruit-bourbon soak.) Set out overnight on porch on a chilly night. Anyone passing by is required to stir them. The next morning, pour everything into cooking pot. Add a pinch of salt. Add water as needed for the fruit to cook down. When done, let it sit for a day for the flavors to meld. Warm and serve with ice cream or process into jars.

I Spy a Michigan Apple Pie

Courtesy of Michigan Apple Committee; original recipe created by Lois Spruytte of Richmond

Pastry Crust Ingredients

2¼ cups pastry flour
1 teaspoon salt
⅔ cup shortening
¼ cup ice water

Mix pastry flour and salt together with a blender. Then add shortening and continue until pea-sized morsels form. Sprinkle water about a tablespoon at a time and toss with a fork until it holds together. Divide into 2 equal portions and shape each into a ball. Flatten each into a disk and wrap separately in plastic or wax paper. Chill dough for about 30 minutes. Roll out first disk and fit into pie plate. Crimp edges and then add prepared filling. Roll out second disk and cover the filled pie. Take a pastry brush and lightly brush half and half over the pie. Sprinkle cinnamon sugar over the top.

Pie Filling Ingredients

8–10 apples (Northern Spy, Paula Red, Empire, Jonathan)
⅓ cup all-purpose flour
1 cup sugar
1¼ teaspoon cinnamon
¼ cup bourbon
scant amount of half and half
sprinkling of extra cinnamon sugar

Peel and slice apples. Take about 3 cups of the apples and place in a saucepan with the bourbon. Cook over medium heat, stirring often until bourbon is diminished and apples are soft. Set aside. Toss remaining apples with the all-purpose flour, sugar and cinnamon. Add the bourbon apples and quickly place in prepared pie pan.

After pie is filled and topped, cover the edges with foil before baking. Place pie on a cookie sheet with sides and bake in a preheated 375-degree oven for 30 minutes. Remove foil and bake for 25 to 30 minutes longer.

THE INTOXICATING APPLE

Top with Michigan Maple Whipped Cream: combine 1 cup heavy whip cream with ¼ cup Michigan maple syrup and a ¼ teaspoon of cinnamon. Place heavy whip in a well-chilled mixer bowl with a well-chilled wire whip beater. Beat on medium speed until cream begins to thicken, then on high speed until soft peaks form. Gradually beat in syrup and cinnamon, scraping the bowl once until blended and stiff peaks form. Refrigerate until needed.

Epilogue
PLANT IT FORWARD

An exciting development in downtown Detroit is the planting of hundreds of apple trees in Palmer Park and the planting of apple trees in compact urban gardens. In the 1880s, Senator Thomas Palmer, who had inherited land and orchards from his grandfather James Witherell, a Supreme Court judge of the Michigan Territory, donated land to the city of Detroit for a park. Until the 1930s, the park had been surrounded by educational farms. In more recent times, the open park space beckoned the creative minds of the city to replant orchards to honor Detroit's heritage.

"The trees were planted for educational and recreational purposes with the hope that we can plant more to establish the industry once again in the city," said Daniel Scarsella, co-owner of Motor City Brewing Works and founding member of People for Palmer Park. Keeping up with statewide trends to plant varieties that have proven to grow well in Michigan's climate, the volunteers planted Jonagold, Idared, Mutsu, Honeycrisp and McIntosh.

In the north, Preserve Historic Sleeping Bear is working with the Sleeping Bear Dunes National Lakeshore National Park Service to preserve older orchards of the region, including on North Manitou Island. The efforts include identifying apple varieties, pruning and harvesting to preserve historic apples and the rich histories of the farmsteads.

Throughout the state, multigenerational farms are planting more apple varieties, offering farm experiences and hosting orchard markets. When you visit, try newly released varieties and apples firmly connected to Michigan's

Epilogue

heritage. Sip the warm, sweet cider of September and October and store jugs in your freezer for wintertime. Go into the orchards to pick a bushel of apples. Toast to your health with a mug of hard cider. Visit the Ridge to see where millions of trees are tucked in the state. Look for apple trees in downtown Detroit.

The rebirth of a community, preservation of heritage farmland and diversification of multigenerational family farms give great hope for Michigan's future. Inspired by the Michigan apple story, I know it is time to plant it forward like those before me. It is time to dig a hole and plant my new apple tree.

Acknowledgements

As a newbie to the apple industry, I am honored to shed light on the impressive history and fresh tastes of the apple. I hope this book inspires spin-offs, as there is still much to be told. Forgive me for mistakes, oversights and limited time and space to include 850 apple farmers. My excursions were often day trips or overnight jaunts that I could accomplish from homes of friends and family. The stories I heard and the history I unearthed traveling five thousand miles were truly remarkable. What amazes me is that there are still hundreds of stories to be told.

This book would not have been written without the encouragement of my husband, Kris. This is my third agricultural book. It is clear that I have a passion for Michigan, and he has always been supportive. My daughters, Julia and Makayla, took it in stride that my attentiveness this year often involved apples.

Thank you, Mom and Dad (Margaret and Richard Martin), for always cheering me on and for building our home by the apple tree and lakes.

A special thank-you to my aunt Jane Schneider of Grand Rapids, who mentioned a young German man who lived on the Ridge. Her prompt led me to investigate the Ridge further, and from there, the book wrote itself.

Thank you to Greg Dumais of The History Press, who invited me to tell the story of Michigan apples.

I was fortunate to have a great extended support team. Amazing Tracie Faupel offered encouragement, research tidbits, recipe organization, limitless carpool rides and a deadline-week survival kit. Ana, thank you for

Acknowledgements

the home stretch goodie bag. Heather, your words of encouragement were just what I needed.

Sharon Steffens, awesome lady of the Ridge, readily agreed to write the foreword. She wrote the perfect message and offered positive energy and support throughout the journey.

Diane Smith of the Michigan Apple Committee verified apple facts, offered insight and shared images and recipes. Alicia Robinette generously shared family photographs. Julia Baehre Rothwell shared her passion for the industry and personal experiences.

Thank you to the many historical societies, archives, libraries and museums for aiding me with my research: Alpine Township Historical Commission, Bentley Historical Library, Benzie Area Historical Museum, Burton Historical Collection, Canadian Museum of History, Central Upper Peninsula and Northern Michigan University Archives, Chelsea District Library, Clarke Historical Library, Fort St. Joseph Museum, Grand Rapids Historical Commission, Grosse Pointe Historical Society, Historical Society of Michigan, the Jewish Historical Society of Michigan, Liberty Hyde Bailey Museum, Michigan Historical Center–Archives of Michigan, Michigan State University, Preserve Historic Sleeping Bear, Rochester Hills Museum at Van Hoosen Farm and William L. Clements Library.

Pat Cederholm of the Alpine Township Historical Commission gets a big shout-out for an abundance of insight, historic photographs, recipes and delicious nutty fruit bars. Thank you to Suanne Shoemaker, Margo Klug, Amy Jo Dayton and King Orchards for your delicious recipes.

Thank you to Steve Elzinga of Erie Orchards for his book *Farming and Flourishing on Sixty Acres*, Dan Young of Tandem Ciders for cider book loans and Gordie Moeller for opening the door to Jack Brown Produce.

Farmers and industry experts answered my many questions and gave generously of their time. In particular: Phil Schwallier and Dr. Nikki Rothwell of Michigan State University and David Bedford of University of Minnesota provided research expertise. (They reviewed sections that are correct. Any mistakes are all mine.) Jim Laubach met me in the orchard so I could experience IPM firsthand. Mike Beck of Uncle John's Cider Mill and Fruit House Winery offered cider expertise. Rob Steffens of Steffens Orchards provided a bird's-eye view of a semi-load of Galas, and Betsy King connected me with two experts and rallied the King Orchards' crew together for an amazing group photo during cherry harvest.

Thank you to the delightful Ward family of Eastman's Antique Apples. Your generous bounty of colorful and unique-tasting historic apples made

Acknowledgements

the prettiest pictures. (More photographs of the apples will be featured on my blog.)

Toward the end of my journey, Peter Hatch provided valuable expertise and pointed me to U.P. Hedrick, Thomas Jefferson and Meriwether Lewis.

Travel expenses were kept to a minimum, thanks to Kami and Bill Kennedy in Grand Rapids and Lorri and John Hathaway in Traverse City. Their homes and company were welcome respites after my travels.

I am so grateful to my dynamic editing team: Leslie Lewis; Phil Lewis, PhD; Sara Wedell; Monica Monsma; and Pam Powell. Thank you for catching my mishaps. You are all brilliant.

Lorri Hathaway, thank you for your insightful edits, creative ideas and celebration box. Even though I did not open it before send-off, I know it was just what I needed.

Appendix
GLOSSARY AND RESOURCES

GLOSSARY

adze: A hand tool with an arched blade used for shaping wood.
antique: Sometimes used to describe apples handed down from generation to generation.
applejack: Strong, concentrated liquor made by freezing hard cider outside in winter.
arpent: One French arpent equals .8448 acres.
auger: A tool that bores holes in wood.
bateau(x): Flat-bottom row boat(s) often elongated and used to haul cargo.
bin: Large wooden or plastic boxes that hold eighteen to twenty bushels of apples.
brix: Sugar content levels of fruit, including apples and grapes.
bushel: A bushel is equal to about forty-eight pounds. There are four pecks in one bushel.
ciderkin: Fermented cider that has been diluted with water.
Controlled Atmosphere Storage (CA): Freshly harvested apples are stored in bins that are stacked and sealed in enormous airtight CA rooms. Once the door is sealed, oxygen is depleted down to just under 2 percent. Temperature, nitrogen and carbon dioxide are set by apple variety and monitored to ensure apples are in "sleep" mode. When apples are needed for the fresh market, the door is opened. With full oxygen restored, apples awaken and taste and appear as if just picked from the orchard.

Appendix

cultivar: A variety that has been selected for breeding and replicated. Cultivated variety.

cuttings: New shoots, or small branches, that have been cut from the tree for the purpose of grafting onto new rootstock to clone the variety from which it originally grew.

cyser: Fermented honey and apple cider. It is a type of mead, which is fermented honey.

détroit: In French, this word means "strait."

dugouts: Canoes made by burning and carving out the center of logs for use on waterways.

eau de vie/eaux de vie: Translated, it means "water of life." Whole fruit is crushed, fermented and distilled. A second distillation creates a clear fruit brandy enjoyed as an aperitif or in culinary dishes.

Fruit Belt: Since the mid-1800s, the region from Berrien County to Antrim County has been deemed the Fruit Belt. It is a region that benefits from tempering climate due to the westerly winds of Lake Michigan. The region's clay loam and sandy loam soils aid in fruit production.

Fruit Ridge: A 158-square-mile region within the Fruit Belt that is 8 miles wide and 20 miles long. It is a land ridge of rolling hills that starts 20 miles north of downtown Grand Rapids and encompasses seven townships and portions of four counties. Within 25 miles of Lake Michigan, the region grows a bounty of fruits and vegetables. Roughly 65 percent of Michigan's apples grow within this region.

heritage apples: Apples that connect to a region as a seedling or as a widely cultivated variety of a particular era.

Homestead Act: Enacted in 1862, this act allowed heads of households to claim 160 acres of land for a nominal fee with the requirement to improve lands within five years of receiving a land grant.

Integrated Pest Management (IPM): Used to manage beneficial and destructive pests in the orchards. Conventional and organic growers rely on IPM to determine best orchard practices from week to week. IPM scouts visit orchards and make field observations and notes. They check traps (small, triangular sticky traps) that attract insects to determine if destructive pests are present. If pest sightings are under the economic threshold, there is no need to spray. If a destructive pest or disease is above this threshold, techniques specific to those issues are instigated.

keeper/keeping apple: An apple that can be kept in cool storage for an extended time into winter.

Appendix

land patent: In 1796, the U.S. Land Grant Office was established to sell federal land to homesteaders in new territories. The patent was a document, or deed, that indicated private landownership.

New France: France established settlements in the New World, the first of which was Quebec in 1608. As explorations continued along the Great Lakes waterways, Michigan became part of the New France territory and remained so until 1763.

oak openings: Prairies surrounded by oak trees and forest.

Peach Ridge: Now called the Fruit Ridge or simply "the Ridge" (see Fruit Ridge). A group of growers established the Peach Ridge Fruit Growers Association in 1928 to promote the region's peach crop. In 1950, the Peach Ridge group planned the first of many Apple Smorgasbords to promote the region's apples and use of apples in various dishes.

peck: Two gallons or eight dry quarts. Apples are often sold in peck or half-peck bags.

pheromone disruption: Thin bands placed strategically in the orchard at the start of the season release an abundance of synthetic pheromone to replicate the female codling moth's scent to attract a mate. The males are disoriented by "the pervasive pheromone levels and cannot find a real calling female in which to mate." This technique interrupts the reproductive cycle before damage is done.

pomace: The seeds, skin, stem and pulp of an apple after it is crushed and pressed into juice.

pomme: apple.

Pomme Caille: Mystery apple, possibly Bourassa, which was a crossbreed of a French variety and likely a crab apple native to Montréal.

pomology/pomological: The study of fruit cultivation.

the Ridge: Initially called the Peach Ridge in 1928. Over time, it was renamed Fruit Ridge.

seedling: An apple tree grown from seed.

seigneurs: Military and aristocrat settlers.

tannin: "Non-volatile phenolic substances" or, in more layperson's terms, the natural texture or body of fruit, such as in apples or grapes.

terroir: In French, it means "earth" or "soil." Ecological and cultural influences of a region, consisting of elements like climate, soil and people. Unique characteristics of a region that impact the taste. Most widely used in relation to wine.

scion or scion wood: A new shoot, or small branch, of a tree that has been cut from the tree for the purpose of grafting onto new rootstock to clone the variety from which it originally grew.

voyageurs: Frenchmen who traveled by canoe into the Great Lakes region and beyond to transport trade goods, particularly fur.

yooper: Nickname for someone who lives in Michigan's Upper Peninsula.

Resources for Maps, Farms and Markets

Fruit Ridge Country Market Guide (www.fruitridgemarket.com)
Great Lakes Cider and Perry (www.greatlakescider.com)
Kent Harvest Trails (www.kentharvesttrails.org)
Michigan Apple Committee (www.michiganapples.com)
Michigan Farmers Market Association (www.mifma.org)
Michigan Land Use Institute (www.mlui.org)
Michigan Wine Industry (www.michiganwines.com)
Real Time Farms (www.realtimefarms.com)

Tree Resources

For the backyard, order trees from Grandpa's Orchard in Coloma (www.grandpasorchard.com) or Southmeadow Fruit Gardens (www.southmeadowfruitgardens.com) in Baroda. In season, find trees at Lehman's Orchard in Niles or Tree-Mendus Fruit farm in Eau Claire.

WORKS CITED

Information was collected during personal interviews with sources. Communication occurred in person and via telephone and e-mail between December 2013 and October 2014 unless otherwise noted herein.

INTRODUCTION

Brilliant Life Quotes. "Mark Twain Quotes." www.brilliantlifequotes.com/inspirational/mark-twain.
Hatch, Peter J. *The Fruits and Fruit Trees of Monticello*. Charlottesville: University of Virginia Press, 1998.
Kerrigan, William. *Johnny Appleseed and the American Orchard*. Baltimore, MD: Johns Hopkins University Press, 2012.

VERITABLE PARADISE OF FRUITS AND MARKETS BLOOM

Apple Journal. http://applejournal.com.
Armstrong, William John. "Berrien County's Great Peach Boom!" http://www.michiganpeach.org/michpeachhistory.html.
Baxter, Albert. *History of the City of Grand Rapids*. Grand Rapids, MI: Munsell & Co., 1891.

Works Cited

Beaudry, Randy. "Apple Maturity in 2014 Is Slower than Expected." Michigan State University Extension. http://msue.anr.msu.edu/news/apple_maturity_in_2014_is_slower_than_expected.
Bersey, John, comp. *Cyclopedia of Michigan: Historical and Biographical, Comprising a Synopsis of General History of the State and Biographical Sketches of Men*. Detroit: Western Publishing and Engraving Co., 1900.
Browne, Ray B., and Lawrence A. Kreiser Jr. *The Civil War and Reconstruction*. Westport, CT: Greenwood Press, 2003.
Bryant, William Cullen. "Ode for an Agricultural Celebration." In *Poetical Works of William Cullen Bryant*. New York: D. Appleton & Company, 1884.
Burton, Clarence M. *Cadillac's Village; or, Detroit Under Cadillac, with a List of Property Owners and a History of the Settlement from 1701 to 1710*. Detroit, 1896.
Burton, Clarence M., William Stocking and Gordon K. Miller, eds. *The City of Detroit, Michigan, 1701–1922*. Vol. 1. Chicago: S.J. Clarke Publishing Company, 1922.
———. *A Sketch of the Life of Antoine de la Mothe Cadillac, the Founder of Detroit*. Detroit: Wilton-Smith Company, 1895.
Canadian Museum of History. "New France." http://www.historymuseum.ca/virtual-museum-of-new-france.
"Canned Food Industry." *Canning Age* 3 (1922).
Carson, Sue. "Research Station 75 Years Old." *News-Palladium*, 1964.
Central Upper Peninsula and Northern Michigan University Archives. "Sam M. Cohodas Papers." http://www.nmu.edu/sites/DrupalArchives/files/UserFiles//MSS-018.html.
Clarke Historical Library. "1702 Antoine Laumet De Lemothe Cadillac." https://www.cmich.edu/library/clarke/ResearchResources/Michigan_Material_Local/Detroit_Pre_statehood_Descriptions/Entries_by_Date/Pages/1702-Antoine-Laumet-De-Lemothe-Cadillac.aspx.
Cohill, Pat. "Farm Wives Catch a Fervor." *Grand Rapids Press*, January 30, 1977.
Coolidge, Judge Orville W. *A Twentieth-Century History of Berrien County, Michigan*. Chicago: Lewis Publishing Company, 1906.
Cunningham, Jeffrey. "Ridge Pioneer Reflects on Spending Time with Mitt Romney's Dad." Mlive. http://www.mlive.com/sparta/index.ssf/2012/02/ridge_pioneer_reflects_on_spen.html.
Da Yoopers. "Da Yoopers Hall of Fame: Sam Cohodas." http://dayoopers.com/fame3.html.
Detroit News. "My Michigan—French Pear Trees." 1939. Article provided by the Grosse Pointe Historical Society.
Discovering Lewis & Clark. http://lewis-clark.org.
Dority, Michael. "Apple Diversity: Paradise Lost." *Repast* 25, no. 3 (Fall 2009).

Works Cited

Downing, Andrew Jackson, and Charles Downing. *The Fruits and Fruit Trees of America*. New York: John Wiley & Son, 1872.

Ergas, Aimée. "The Upper Peninsula's Cohodas Family." *Michigan Jewish History* 42 (Fall 2002).

Farmer, Silas. *The History of Detroit and Michigan; or, the Metropolis Illustrated; a Chronological Cyclopaedia of the Past and Present, Including a Full Record of Territorial Days in Michigan, and the Annals of Wayne County and Early Michigan*. Detroit: Silas Farmer and Co., 1884.

Fort St. Joseph Museum. Interpretive Exhibits. Niles, MI, September 13, 2014.

Fruit Growers News. "IFTA Helped Growers Bring Fruit Trees Down to Size." http://fruitgrowersnews.com/index.php/magazine/article/IFTA-helped-growers-bring-fruit-trees-down-to-size.

Gerber. "Gerber's History & Heritage." http://gerber.com/our-story/our-history-and-heritage.

Hacker, David. "Meet Mr. Sam...a Real Yooper." *Ludington Daily News*, 1987.

Hall, T.P., and S. Farmer. *Grosse Pointe on Lake Sainte Claire: Historical and Descriptive*. Detroit: Silas Farmer and Co., 1886. Republished by Gale Research Company, Detroit, 1974.

Hampton, Charles F. *Michigan Log Cabins and Hard Cider*. Brighton, MI: Green Oak Press, 1984.

Hartman, S.B., and H.G. Eustance. "Can the General Farmer Afford to Grow Apples?" Michigan State Agricultural College Experiment Station, 1909.

Hatch, Peter J. *The Fruits and Fruit Trees of Monticello*. Charlottesville: University of Virginia Press, 1998.

Hedrick, U.P. *A History of Horticulture in America to 1860*. New York: Oxford University Press, 1950.

———. *The Land of the Crooked Tree*. New York: Oxford University Press, 1948.

Holmes, Ken. "Remember When...with Ken." Alberta Report, 2011.

Judge, Arthur I., ed. "A History of the Canning Industry by Its Most Prominent Men." *The Canning Trade* 37, no. 21 (1914).

Juniper, Barrie E., and David J. Mabberley. *The Story of the Apple*. Portland, OR: Timber Press, 2006.

Kent, Timothy J. *Fort Pontchartrain at Detroit*. Vols. 1–2. Ossineke, MI: Silver Fox Enterprises, 2001.

Kraft, Merlin. "The Henry Kraft and Son Apple Storage." Alpine Township Historical Commission, 2013.

Lanman, Charles. 1867. *The Life of William Woodbridge*. Washington, D.C.: Blanchet & Mohun, 1867.

Laug, Cindy. "Women on the Ridge." Grand Rapids Historical Commission. http://www.historygrandrapids.org/photoessay/1691/women-on-the-ridge.

WORKS CITED

Leeson, M.A. *History of Kent County, Michigan: Together with Sketches of Its Cities, Villages and Townships and Biographies of Representative Citizens*. Chicago: Chas C. Chapman and Co., 1881.

Lehnert, R. "About Phil Brown Welding Corp." http://www.philbrownwelding.com/about-phil-brown-welding-corp.

Liberty Hyde Bailey Museum. http://libertyhydebailey.org.

Lyon, T.T. *History of Michigan Horticulture*. Lansing, MI: Thorp & Godfrey, 1887.

MacDonald, Eric, and Arnold R. Alanen. *Tending a "Comfortable Wilderness": A History of Agricultural Landscapes on North Manitou Island, Sleeping Bear Dunes National Lakeshore, Michigan*. Omaha, NB: U.S. Department of Interior, National Park Service, Midwest Field Office, 2002.

Michigan Apple Committee. http://www.michiganapples.com.

Michigan Department of Environmental Quality. "Great Lakes Map." http://www.michigan.gov/deq/0,4561,7-135-3313_3677-15926--,00.html.

Michigan Legislature. "Michigan History." http://www.legislature.mi.gov/documents/publications/manual/2001-2002/2001-mm-0003-0026-History.pdf.

Michigan Manufacturer and Financial Record 15, no. 7 (1915): 4.

Michigan State Board of Agriculture. *Ninth Annual Report of the State Board of Agriculture of the State of Michigan*. Lansing: W.S. George and Co., 1870.

———. *Twelfth Annual Report of the Secretary of the State Board of Agriculture*. Lansing: W.S. George and Co., 1875.

Michigan State Historical Society. *Collections and Researches Made by the Michigan Pioneer and Historical Society*. Vol. 32. Lansing: Robert Smith Printing Company, 1902.

———. *Collections and Researches Made by the Michigan Pioneer and Historical Society*. Vol. 35. Lansing: Wynkoop Hallenbeck, 1907.

———. *Collections and Researches Made by the Michigan Pioneer and Historical Society*. Vol. 38. Lansing: Wynkoop Hallenbeck, 1912.

———. *Report of the Pioneer Society of the State of Michigan*. Vol. 3. Lansing: W.S. George and Co., 1879.

———. *Report of the Pioneer Society of the State of Michigan*. Vol. 5. Lansing: W.S. George and Company, State Printers and Binders, 1884.

Michigan State Horticultural Society. *Annual Report of the State Horticultural Society of Michigan*. Lansing: W.S. George and Co., 1873.

———. *Annual Report of the State Horticultural Society of Michigan*. Vol. 16. Lansing: Thorp and Godfrey, 1886.

———. *Annual Report of the State Horticultural Society of Michigan*. Vol. 17. Lansing: Thorp and Godfrey, 1887.

———. *Annual Report of the State Horticultural Society of Michigan*. Vol. 38. Lansing: Wynkoop, Hallenback, Crawford Co., 1909.

Works Cited

———. *Third Annual Report of the State Horticultural Society of Michigan*. Lansing: W.S. George and Co., 1874.
Michigan State Pomological Society. *Annual Report of the State Pomological Society of Michigan*. Vol. 1. Lansing: W.S. George and Co., 1872.
———. *Annual Report of the Secretary of the State Pomological Society of Michigan*. Lansing: W.S. George and Co., 1875.
———. *Annual Report of the Secretary of the State Pomological Society of Michigan*. Lansing: W.S. George and Co., 1877.
———. *Second Annual Report of the Michigan State Pomological Society*. Lansing: W.S. George and Co., 1873.
Milan, Jon. *Old Chicago Road: US-12 from Detroit to Chicago*. Charleston, SC: Arcadia Publishing, 2011.
Nabhan, Gary Paul, ed. "Forgotten Fruits Manual & Manifesto: Apples." Renewing America's Food Traditions (RAFT) Alliance. http://garynabhan.com/pbf-pdf/applebklet_web-3-11.pdf.
National Constitution Center. "American Spirits: The Rise and Fall of Prohibition." Exhibited curated by Daniel Okrent. http://prohibition.constitutioncenter.org.
National Park Service. "D.H. Day." http://www.nps.gov/slbe/historyculture/dkhday.htm.
Paasch, Kathryn. *Sparta Township*. Charleston, SC: Arcadia Publishing, 2011.
Palmer, Friend. *Early Days in Detroit*. Detroit: Hunt and Jones, 1906.
———. "Early Days in Grosse Pointe." In *Tonnancour*. Vol. 2. Grosse Pointe Farms, MI: Tonnancour Associates, 1997. http://www.gphistorical.org/pdf-files/tonnancour/friend.pdf.
"The Peach Ridge Story." Alpine Township Historical Commission, 1971.
Peppel, Fred. "Underground Railroad in Kalamazoo." http://www.kpl.gov/local-history/general/underground-railroad-old.aspx.
Pere Marquette Railway. "One Million Acres of Lands in the State of Michigan for Sale by the Grand Rapids and Indiana Railroad Company." 1870s brochure. William L. Clements Library, Ann Arbor, MI.
Portrait and Biographical Record of Kalamazoo, Allegan and Van Buren Counties, Michigan. Chicago: Chapman Bros., 1892.
Preserve Historic Sleeping Bear. http://www.phsb.org.
Ray's Place. "History of Clyde Township, MI." http://history.rays-place.com/mi/alle-clyde.htm.
Reynolds, D.B. "Early Land Claims in Michigan." Michigan Department of Conservation, 1940.
Rochester Era. "Avon in Height of Season Yields Variety of Apples." 1931.

Works Cited

Rothwell, Julia Baehre. "If Not Now, When?" U.S. Apple Association Address, Chicago, 2011.

Saline Area Historical Society. "Saline Railroad Depot Museum." http://salinehistory.org/index.php?section=sites&content=railroad_depot.

Scott, J. Amadeaus. "These Branches Still Bear Fruit." *Repast* 25, no. 3 (Fall 2009): 6–7.

Sentinel-Leader. September 3, 1958.

Southwest Michigan Business and Tourism Directory. "The History of Watervliet, Michigan." http://www.swmidirectory.org/History_of_Watervliet.html.

Sparta Prisoner of War Camp. Alpine Township Historical Commission.

Sprague, Elvin L., and Mrs. George N. Smith. *Sprague's History of Grand Traverse and Leelanaw Counties, Michigan*. Logansport, IN: B.F. Bowen, 1903.

Stark, George W. "We Old Timers...and Apples, Too." *Detroit News*, 1941. Article provided by the Grosse Pointe Historical Society.

Stroup, Donald. *The Manistee and Northeastern: The Life and Death of a Railroad*. Lansing: Historical Society of Michigan, 1964.

University of Illinois Extension. "Apple Facts." http://urbanext.illinois.edu/apples/facts.cfm.

U.S. Department of Agriculture. "Apples: Michigan Fruit Inventory, 2000–2001." http://www.nass.usda.gov/Statistics_by_State/Michigan/Publications/Michigan_Rotational_Surveys/mi_fruit01/apples.pdf.

Vogt, Pat. "Sharon Steffens: 'Getting Agri-Power.'" *Michigan Farmer*, 1977.

Watson, Ben. *Cider Hard and Sweet*. Woodstock, VT: The Countryman Press, 2009.

Watson, Jeannie. "Pere Claude Jean Allouez & Jesuit Mission de Saint Joseph. Fort St. Joseph." http://berrien.migenweb.net/Profiles/Watson/BIO30ClaudeAllouez.pdf.

Weeks, George. *Sleeping Bear: Yesterday & Today*. Ann Arbor: University of Michigan Press, 1990.

Weld, Isaac, Jr. *Travels Through the States of North America, and the Provinces of Upper and Lower Canada During the Years 1795, 1796 and 1797*. 2nd ed. London: John Stockdale, Piccadilly, 1799.

Wilder, Marshall P. "Address at the Twentieth Session of the American Pomological Society." Boston: American Pomological Society, 1885.

Withers, James. "Commercial Fruit Evaporating." *American Gardening* 14 (1893): 339.

WORKS CITED

APPLES TO APPLES

Cornell University. "Apple Breeding." http://www.fruit.cornell.edu.
Fruit Growers News. "IFTA Helped Growers Bring Fruit Trees Down to Size." http://fruitgrowersnews.com/index.php/magazine/article/IFTA-helped-growers-bring-fruit-trees-down-to-size.
Jack Brown Produce. http://www.jackbrownproduce.com.
Lonie, Iaine M. *The Hippocratic Treatises "On Generation," "On the Nature of the Child," "Diseases IV."* Berlin: Walter D. Gruyter & Co., 1981.
Silverton, Jonathan. *An Orchard Invisible: A Natural History of Seeds*. Chicago: University of Chicago Press, 2010.
Southwick, Larry. *Grafting Fruit Trees*. North Adams, MA: Storey Publishing, 1981.
Thoreau, Henry David. "Wild Apples." *Atlantic Monthly*, 1862.
Zhang, Sarah. "Awesome Apple Vintage Art." http://www.motherjones.com/environment/2013/04/old-apples-of-new-york-book-vintage-illustrations.

FOUR SEASONS

Bailey, Liberty Hyde, Jr. *The Apple-Tree*. New York: Macmillan Company, 1922.
Cameron, Layne, Rufus Isaacs and Jennifer Martin. "USDA Awards MSU $6.9 Million Grant to Help Bees." Michigan State University Today. http://msutoday.msu.edu/news/2014/usda-awards-msu-69-million-grant-to-help-bees.
Collett, Lloyd. "About the Apple: *Malus domestica*." Oregon State University Extension Services. http://extension.oregonstate.edu/lincoln/sites/default/files/about_the_apple.lc_.2011.pdf.
Frazier, M. 1997. "A Short History of Pest Management." Penn State Extension. http://extension.psu.edu/pests/ipm/schools-childcare/schools/educators/curriculum/contents/shorthistory.
Hagman, Arthur A., ed. *Oakland County Book of History: The Sesquicentennial Publication, 1820–1970*. Oakland County, Michigan Sesquicentennial Committee, 1970.
Irish-Brown, Amy. "Orchard Observations." Michigan Apples Committee Facebook page, 2014.
Michigan Department of Agriculture and Rural Development. "Migrant Labor Housing." http://www.michigan.gov/mdard/0,4610,7-125-1569_45168—,00.html.
Milkovich, Matt. "McDonald's TV Commercials Promote Michigan Apples." Fruit Grower News. http://fruitgrowersnews.com/index.php/magazine/article/mcdonalds-tv-commercials-promote-michigan-apples.

WORKS CITED

A Virginia Farmer. *Roman Farm Management: The Treatises of Cato and Varro Done into English, with Notes of Modern Instances.* New York: Macmillan Company, 1918.

INTO THE ORCHARDS

Aussie Apples. "Granny Smith." http://www.aussieapples.com.au/aussie-grown-varieties/granny-smith.aspx.
Bailey, Liberty Hyde, Jr. *The Apple-Tree.* New York: Macmillan Company, 1922.
Hedrick, U.P. *A History of Horticulture in America to 1860.* New York: Oxford University Press, 1950.
King Orchards. http://kingorchards.com.
Michigan Apple Committee. http://www.michiganapples.com.
Michigan State Pomological Society. *Annual Report of the State Pomological Society of Michigan.* Lansing: W.S. George & Co., 1874.
New York Apple Country. "Apple Varieties of New York State." http://www.nyapplecountry.com/varieties.
Nye's Apple Barn. http://www.nyesapplebarn.com.
Purdue University–Rutgers University–University of Illinois Apple Breeding Program. http://www.hort.purdue.edu/newcrop/pri/.
Urban, Sylvanus. *The Gentleman's Magazine.* London: John Nichols, 1791.

TASTES OF HISTORY

Burford, Tom. *Apples of North America.* Portland, OR: Timber Press, 2013.
Cunningham, Jeffrey. "Ridge Pioneer Reflects on Spending Time with Mitt Romney's Dad." Mlive. http://www.mlive.com/sparta/index.ssf/2012/02/ridge_pioneer_reflects_on_spen.html.
Dennis, F.G., Jr., G.M. Kessler and H. Davidson, H. *From Seed to Fruit: 150 Years of Horticulture at Michigan State, 1855–2005.* Michigan State University Department of Horticulture. East Lansing, MI: University Printing, 2007.
Eastman's Antique Apples. "Apple Varieties: Borsdörffer." http://www.eastmansantiqueapples.com/varieties/varietiesab.html.
Folger, John Clifford, and Samuel Mable Thomson. *The Commercial Apple Industry of North America.* New York: Macmillan Company, 1921.

WORKS CITED

Henderson, Sandy. "Forgotten Fruits." *Chicago Reader*, October 19, 1989. http://www.chicagoreader.com/chicago/forgotten-fruits/Content?oid=874591.
King Orchards. http://kingorchards.com.
Lyon-Jenness, Cheryl. *For Shade and for Comfort: Democratizing Horticulture in the Nineteenth-Century Midwest*. West Lafayette, IN: Purdue University Press, 2004.
Michigan Apple Committee. http://www.michiganapples.com.
Michigan State Horticultural Society. *Fifteenth Annual Report of the Secretary of the State Horticultural Society of Michigan*. Lansing: Thorp & Godfrey, 1886.
Michigan State Pomological Society. *Annual Report of the State Pomological Society of Michigan*. Vol. 2. Lansing: W.S. George & Co., 1873.
———. *Annual Report of the State Pomological Society of Michigan*. Vol. 7. Lansing: W.S. George & Co., 1877.
New England Apple Association. http://www.newenglandapples.org.
New York Apple Association. http://www.nyapplecountry.com.
Thompson, Tad. "Michigan's Gained Time Applied to Future of Jack Brown and Others." *Produce News*, January 20, 2014.
Thoreau, Henry David. "Wild Apples." *Atlantic Monthly*, 1862.
Tree-Mendus Fruit Farm. "Tree-Mendus Fruit Farm History." http://treemendus-fruit.com/album_2_015.htm.
Warder, J.A. *American Pomology*. New York: Orange Judd & Co., 1867.
Weld, Isaac, Jr. *Travels Through the States of North America, and the Provinces of Upper and Lower Canada During the Years 1795, 1796 and 1797*. 2nd ed. London: Piccadilly, 1799.

THE HUNT FOR SWEET OCTOBER

Bailey-Boorsma, Joanne. "Alpine Township Apple Grower Achieves Rare Title of Master Cider-Maker." Mlive. http://www.mlive.com/northwestadvance/index.ssf/2010/12/alpine_township_apple_grower_a.html.
Cammel, Ron. "New Champ in Michigan Apple Cider Making Contest." Mlive. http://www.mlive.com/business/index.ssf/2011/12/new_champ_in_michigan_apple_ci.html.
Donahue, Denise. "Hill Brothers Earn State's Most Distinguished Cider-Making Award." MichiganApples.com. http://www.michiganapples.com/News/Archive/ID/320/Hill-Brothers-earn-states-most-distinguished-cider-making-award.

WORKS CITED

Historic Bowens Mills. "The Past Lives Again at Historic Bowens Mills." http://www.bowensmills.com/Historic%20Bowens%20Mills%20Historical%20Park/HISTORY/old%20cider%20grist%20mill%20history%201864.htm.
Massachusetts Department of Agricultural Resources. http://www.mass.gov/eea/agencies/agr.

THE INTOXICATING APPLE

Beneson, Bob. "One Professor's Spirited Enterprise." Symposium Magazine. http://www.symposium-magazine.com/one-professors-spirited-enterprise.
Berglund, Kris. Phone and e-mail communication with author, 2009 and 2010.
Blake Farms. http://blakefarms.com.
Davidhisar, Joanne. "Michigan's Hard Cider Comes Naturally." Michigan State University Extension. http://msue.anr.msu.edu/news/michigans_hard_cider_comes_naturally.
Epicurious. "Moscow Mule." http://www.epicurious.com/articlesguides/drinking/cocktails/moscowmule.
Farmer, Silas. *The History of Detroit and Michigan*. Vol. 1. Detroit: S. Farmer and Company, 1889.
Halfpenny, Rex. "Meckley's Flavor Fruit Farm, Somerset Center." Michigan Beer Guide, 2014.
Juniper, Barrie E., and David J. Mabberley. *The Story of the Apple*. Portland, OR: Timber Press, 2006.
Organic Scrumpy. "J.K.'s Orchard Gate Gold." http://organicscrumpy.com/AlmarFarmhouseCider.html.
Schultz, E.J. "Cider Seen as Next 'Craft' Brew as Sales Climb 25% This Year." *Advertising Age*, November 28, 2011. http://adage.com/article/news/cider-craft-brew-sales-climb/231198.
Southmeadow Fruit Gardens. "Online Catalog: Cider Apples." http://southmeadowfruitgardens.com/FruitTreeCatalog.html.

EPILOGUE

People for Palmer Park. http://peopleforpalmerpark.org.

WORKS CITED

GLOSSARY

United States Department of the Interior, Bureau of Land Management. "The Official Federal Land Records Site." http://glorecords.blm.gov/default.aspx.

The Wittenham Hill Cider Portal. "Tannin in Cider Apples." www.cider.org.uk/tannin.htm.

Index

A

Allouez, Pere Claude Jean 17
Almar Orchards 27, 96, 101, 110
Alpine Township Historical
  Commission 13, 53, 54, 71
Alt 23
American Agri-Women 51
American Horticultural Society 45
Anderson 23
Antrim County 63
apple brandy 113
Apple Charlie's Orchard and Cider
  Mill 105
applejack 17, 20, 133
Apple Keg 29
Apple King 28
apple pie 85, 124
AppleRidge 65
Appleseed, Johnny 9, 12, 20, 54, 117
Apple Smorgasbord 46
Arends, Luke 10, 49, 80
Armock 23
Armstrong, John 18
Aveneau, Claude 17
Avon 22

B

Baehre 23
Bailey, Liberty Hyde, Jr. 23, 58, 63, 78
Bakker, Al 67
Bakker's Acres 67, 83
Baldwin 12, 20, 26, 31, 37, 39, 40, 94,
  106, 117
Bardenhagen, Jim 80
Bardenhangen Farms 67
Beck, John 50
Beck, Mike 103, 114, 115, 123
Bedford, David 59, 60, 82
BelleHarvest–Belding Fruit 65
Ben Davis 27, 37, 40
Benton Harbor Market 28
Benzie County 25
Berglund, Kris 113
Berrien County 34, 63, 134
Beuham, Frederic 27
Big Dan's U-Pick'em and Farm Market
  69, 75, 105, 111
Bixby Farms 74
Black Star Farms 114
Blake's Hard Cider 121
Blondee 81
Blue Pearmain 90

Index

B. Nektar 123
Borsdörffer 91
Bourassa 17, 93, 135
Brechting 23
Brechting's Farm Market 56, 69, 84
Briggs, Jay 115, 119
Brown 23
Brown, Jack and Aleta 41
Brown, Phil 49
Brownwood Farms 66
Bryant, William Cullen 15
Burnette Foods 65, 66
Burnett, William 21

C

Cadillac, Antoine de le Mothe 93
Calvados 113
Calville Blanc d'Hiver 17, 19, 20, 93
Calville Rouge d'Automne 20, 93
Canada 16, 18, 44, 56, 117
Canfield, Connee 50
canning 39
Cederholm, Pat 13, 71
Chapman, John 12, 20, 117
Charlevoix Applefest 83, 93
Chelsea 28, 101
Chelsea Farmers Market 78
Cherry Central 65
Cherry Growers 65
Cherry Marketing Institute 113
Chestnut Crab 68, 92
Chevalier 17
Chicago 21, 28, 29, 33, 44, 74, 95, 97
Chicago Road 19
chocolate 18
ciderkin 17, 118, 133
Clark, William 18
Cohill, Pat 50
Cohodas Brothers Fruit Company 29
Cohodas, Harry 29
Cohodas, Sam 13, 29
Coloma Frozen Foods 65
Cone, Linus 22
controlled atmosphere storage 49, 133

Corey Lake Orchards 80, 114
Cornell University 52, 58
Country Mill 110
Court Pendu Plat 91
Cox's Orange Pippin 82, 91
Crane's Pie Pantry Restaurant and Bakery 111, 122
Crimson Crisp 82
Crispin 84
Crow, Ben 121
Crow's Hard Cider 121
cyser 113, 114, 123

D

Davis, William 28
Detroit 17, 18, 19, 93, 98, 99, 127
Detroit Red 17, 19, 20, 93
Detroit River 19
Dexter Cider Mill 67, 101, 102, 106
Dietrich 23
Dietrich, Al 66
Dietrich, Helen 71
Dietrich Orchards 71, 106
Doherty, Mark 59
Douglas Valley 30
Drake, Dawn 73
Duchess of Oldenburg 12, 32, 40, 90, 95
Dunneback 23
Dwarf Fruit Tree Association 46

E

Earth First Farms 110
Eastern Market 28
Eastman's Antique Apples 13, 90, 113, 120
eau de vie 114, 134
Ebers 23
Ed Dunneback and Girls 69
Ed Dunneback and Girls Farm Market 61, 74
Emery, Bill 109
Empire 82
Empire Orchards 57
Engelsma, Jim 109

INDEX

Engelsma's Apple Barn Cider Mill 109
Erie Canal 19
Erie Orchards 74
Erwin, Bill 109
Erwin Orchards 109
Esopus Spitzenburg 20, 32, 92, 95, 119, 121

F

Fameuse 12, 17, 20, 22, 32, 37, 44, 56, 78, 82, 92, 93, 96, 97, 120
Farmer, Silas 17
Fearsome Foursome 50
Flint Farmers Market 44
Ford, Gerald 48
Ford, Henry 38, 42
Fort Pontchartrain du Detroit 16
Fort St. Joseph 17, 33
Forty-Five North 119
Fox Barn Market and Winery 122
Fox, Kellie 122
Fox, Todd 66, 122
Franklin, Benjamin 90
Franklin Cider Mill 105, 106
Fremont Canning Company 41
French and Indian War 17
Friske Orchards 74, 101, 111
Fruit Belt 29, 32, 52, 63, 97, 134
Fruitful Orchard and Cider Mill 105
Fruit Grower News 65
fruiting walls 52
Fruit Picking Equipment Company 45
Fruit Ridge 13, 23, 41, 52, 134, 135
Fuji 13, 52, 58, 59, 83, 84
Fulton Street Market 39

G

Gala 52, 82, 89, 109, 114
Gavin Orchards 67
General Land Office of the United States 19
Genesee County 20
Gerber 41, 66, 97
Ginger Gold 12, 67, 74, 80

Golden Delicious 56, 66, 67, 83, 84, 89, 95, 109, 114, 119
Golden Russet 26, 32, 37, 52, 68, 91, 97, 119, 122
Golden Supreme 80
Golden, W. 26
Good Neighbor Organic 121
grafting 57
Grandpa's Cider Mill 105
Grandpa's Orchard 136
Grand Rapids Press 51
Grand River 21, 23
Grand Traverse 25, 26
Grand Traverse Union Agricultural Society 26
Gravenstein 80, 92
Great Depression 44
Great Lakes Cider and Perry Festival 91
Grimes Golden 27
Grosse Ile 17

H

Hagger, Rob 122
Haight, H.R. 26
hard cider 115
Harrison, William 21
Hatch, Peter 19
Heavilin, Eric 122
Hidden Rose 92
Hill 23
Hill Bros. 108
Hill, Jim 13, 108
Hill, Joan 50
Hilltop Orchards and Nurseries 46
Historic Bowens Mill 108
Honeycrisp 10, 12, 51, 58, 59, 60, 67, 74, 82, 83, 104, 121, 127
Hoppin, George 33
Hoxsie's Farm Market 66, 74, 96, 110
Hoyt, Benjamin 21
Hubbard, Dayton 114
Huron 17
Huron Township's Applefest 105
Husted Market 81

Index

Husted, Noah P. 32
Hy's Cider Mill 104

I

Idared 127
Indian Summer 65, 66
Integrated Pest Management 70, 134
International Fruit Tree Association 46

J

Jack Brown Produce 41, 65, 66
Jay Treaty 18
Jefferson, Thomas 12, 18, 19, 95
Jelinek Orchards 83
Jesuit 11, 15, 17, 33
J.K.'s Scrumpy 96, 117
Johnson, Lady Bird 48, 53
Johnson Orchards 27, 66
Jollay Orchards 105
Jonagold 127
Jonathan 27, 40, 84, 89, 92, 95, 109, 114

K

Kalamazoo 21, 22, 24, 81
Kapnick Orchards 78
Keepsake 61
Kent 20
Kent County 21, 22, 34
Kent County Agricultural Society 31
Kilcherman's Christmas Cove Farm 82, 102, 120
King Orchards 13, 35, 71, 78, 84, 93, 96
Klackle Orchards 27
Klein 23
Klein Cider Mill and Market 110
Klenk 23
Koan, Jim 96, 117
Kober 23
Kraft 23

L

Ladd, E.P. 26
Leaman's Green Apple Barn 74

Lecklider, Steve 120
Leelanau Peninsula 25
Left Foot Charley 119
Lehman's Orchards 120, 136
Lehman's Tap House 121
Lenawee County 30
Leo Dietrich and Sons 66
Lesser Farms and Orchard 101
Lewis, Meriwether 18
Linardos, John 117
Linda Mac 58
Lodi 73, 78
Log Cabin and Hard Cider Campaign 21
Long Family Orchard, Farm and Cider Mill 105
Lutes, Lee 114
Lyon, Theodatus Timothy 31

M

Mackey, Dennis 13, 30, 70, 121
Mackey, Kyle 71, 121
Mackinac Island 25
Mackinaw 22
Macoun 82
Maiden's Blush 95
Malus domestica 11
Manistee and Northeastern Railroad 29
Marketing and Bargaining Act of 1973 51
master cider maker 108, 109
McIntosh 12, 49, 56, 96, 109, 127
McIntosh, Bruce 117
McIntosh Orchards 92, 117, 122
Meckleys' Flavor Fruit Farm 121
Michigan Agricultural College 58
Michigan Agricultural Cooperative Marketing Association 73
Michigan Apple Committee 44, 52, 61, 73, 86, 116, 124, 136
Michigan Farmer 50
"Michigan Fever" 20
Michigan State Apple Commission 44
Michigan State Board of Agriculture 31

Index

Michigan State College 23
Michigan State Horticultural Society 12
Michigan State Pomological Society 12, 26, 31, 32, 95, 96, 97
Michigan State University 31, 57, 65, 113
Michigan Territory 19, 21, 94, 127
Michilimackinac 18
Mission de Saint Joseph 17
Moelker Orchards Farm Market and Bakery 27, 109, 111
Mollies Delicious 80
Monroe 17, 22, 32
Monticello 12, 18, 19
Moscow Mule 115
Mother Earth News 71
Motor City Brewing Works 117, 120, 127
MSU Clarksville Research Center 60
Mutsu 52, 67, 84, 85, 127

N

New France 16, 56, 135
Newhall 27
New York 29
New York State Agricultural Experiment Station 59, 78
Niles 17, 33, 120
NJ Fox and Sons 66, 122
Northern Natural 121, 122
Northern Spy 12, 26, 30, 32, 37, 40, 44, 50, 66, 67, 114, 120
North Manitou Island 23, 26, 56, 98, 127
Northwest Territory 18, 19
Nyblad 23
Nye's Apple Barn 85

O

Oakland County 22
Old Mission Peninsula 25
Old Orchard 65
Old Sauk Trail 19
Overhiser, Allan 68
Overhiser Orchards 27, 68, 73, 74, 78, 80

P

Page, Abel 22
Palmer, Friend 22
Palmer, Thomas 127
Parmalee, George 26, 97
Parmenter's Northville Cider Mill 108
Parshallville Cider Mill 106
Paula Red 10, 13, 49, 73, 74, 80, 81
Peach Ridge 135
Peach Ridge Fruit Growers Association 52
Peach Ridge Orchard Supply 45
People for Palmer Park 127
Peterson Farms 65, 66
Phillips, Brian 44, 95
Phillips Orchards 27, 44, 82, 84, 95, 110
Piccard, Jean Rasch 99
Plymouth Orchards and Cider Mill 105, 110
polar vortex 64
Pomme Caille 18, 135
Pomme d'Api 17, 92, 93
Pomme de Neige 17
Pomme Gris 17, 19, 20, 94, 97
Porter's Perfection 91, 96
Preserve Historic Sleeping Bear 127
Pristine 78
Pulcipher, John 26
Purdue University 79, 82, 85

R

Ramsdell, Jonathan 26
Rasch 23
Rasch, Don 45, 70, 71
Rasch Family Orchards 45
Red Astrachan 26, 32, 37, 44, 96
Red Cedar Spirits 115
Red Delicious 56, 59, 83, 89, 97
Redfree 82
Reinette de Misnie 91
Reister 23
Revolutionary War 18
Rhode Island Greening 26, 32, 37, 40, 44, 97, 106, 114, 120, 121

INDEX

ribbon farms 12, 16, 17, 28
Ridgeview Orchards 66
Robinette, Barzilla 39
Robinette, Ed 50, 68
Robinette, Jim 50
Robinette's Apple Haus and Winery 27, 68, 104, 119, 122
Rochester 44
Rome 97
Rome Beauty 66
Rosseau 93
Rothwell, Julia Baehre 41, 49
Rothwell, Nikki 69, 119
Roxbury Russet 20, 31, 46, 97, 119, 121
Royal Farms 122
Ruby Mac 56
Ruesink, Bruce 114
Rutgers University 79, 82

S

Saginaw 32
Saginaw Valley 22
Sansa 83
Scarsella, Dan 117, 127
Schaefer 23
Schaefer, John 41, 65
Schwallier 23
Schwallier, Phil 13, 57, 65, 71
Schwallier's Country Basket 13, 109
Sherwood, Harvey 28
Shiawassee Beauty 13, 23, 27, 44, 93, 97
Sietsema, Andy 119
Sietsema Orchards and Cider Mill 74, 104, 110, 119
Sietsema, Skip 104
Sleeping Bear Dunes National Lakeshore 27, 42, 56, 58, 90, 98, 127
Smeltzer Orchard 27, 65, 66
Smith, Diane 73
Smith, Hezikiah 24
Snow 12, 32, 37, 44, 56, 78, 92, 93
Sonday, Becca 114
Southmeadow Fruit Gardens 90, 136

Spartan 80
Spicer Orchards 101, 106, 122
Stark Bro's Nurseries and Orchards 27, 56, 84
Stark Red Delicious 27
Stark, William 27
Steffens 23
Steffens, John 44
Steffens Orchards 10, 74, 81
Steffens, Sharon 10, 13, 50, 86
St. Joseph 21, 22, 28
St. Lawrence Valley 16, 56, 93
Straits of Mackinac 18
SweeTango 68, 83

T

Tandem Ciders 119, 120
Teichman, Bill 92
Teichman, Herb 92
Thome 44
Thome, Bernard 56
Thome, Harold 23, 44
Thome, JoAnn 13, 50
Thome, Mitch 24
Thome, Steve 24, 66
Thoreau, Henry David 55, 90
Three Cedars Farm 105
Tiffin, Edward 19
Tippecanoe and Tyler Too 22
Transparent 78
Traverse City 66, 70
Tree-Mendus Fruit 92, 136
Tritten, Bob 108
Troy 22
Truesdell, Beebe 23
Twain, Mark 13
Twenty Ounce 20

U

Umlor 23
Umlor, Walter 45
Uncle John's Cider Mill and Fruit House Winery 103, 104, 115
Underground Railroad 24

INDEX

University of Illinois 79, 82
University of Michigan 19, 34
University of Minnesota 52, 60, 80, 82
University of Minnesota Agricultural Experiment Station 59
U.S. Apple Association 51

V

Vander Mill 119
victory gardens 44

W

Wagener 26, 27, 31, 32, 40, 44, 50, 84, 98
Ward 90, 91
Ward, Cindy 91
Ward, Rafe 113
Washington State University 52
Wayne County 19
Wells Orchards 27, 109
Westfield Seek-No-Further 98
Westview Orchards and Adventure Farm 27
Whigs of Detroit 22
White House 48
White Pigeon Land Office 20
White, Variety 20
Wiard and Son 32
Wiard, Phil 33, 50
Wiard's Orchards and Country Fair 27, 50
Wiesen, Dan 57
"Wild Apples" 90
Winesap 114
Winter Banana 106
Witherell, James 127
Wittenbach Orchards 27
Women for the Survival of Agriculture in Michigan 50
Woodbridge, William 20, 28
Woodward, Augustus 19
Woodward Avenue Market 28
World's Columbian Exposition 33

Y

Yates Cider Mill 103, 106
Yellow Bellflower 19, 98
Youngquist 23
Ypsilanti 20, 32

Z

Zestar! 80, 83

About the Author

Sharon Kegerreis is co-author of two previous books on Michigan wine, one of which was a 2008 Michigan Notable Book. She is passionate about Michigan and enjoys outdoor and travel adventures and gardening. Sharon grew up in Charlevoix with a big apple tree in the backyard. For more than a decade, she has lived with her family in Chelsea, where she has been working hard to grow apple trees on her unruly five acres. Get in touch at www.sharonmk.com.

Photo by Kris Kegerreis.

Visit us at
www.historypress.net

This title is also available as an e-book